Eros *and* Evolution

Eros and Evolution

A Natural Philosophy of Sex

Richard E. Michod

§ HELIX BOOKS

▲▲ ADDISON-WESLEY PUBLISHING COMPANY

Reading, Massachusetts ◆ Menlo Park, California ◆ New York
Don Mills, Ontario ◆ Wokingham, England ◆ Amsterdam ◆ Bonn
Sydney ◆ Singapore ◆ Tokyo ◆ Madrid ◆ San Juan
Paris ◆ Seoul ◆ Milan ◆ Mexico City ◆ Taipei

Many of the designations used by manufacturers and sellers to distinguish their products are claimed as trademarks. Where those designations appear in this book and Addison-Wesley was aware of a trademark claim, the designations have been printed in initial capital letters.

Library of Congress Cataloging-in-Publication Data
Michod, Richard E.
 Eros and evolution : a natural philosophy of sex / Richard E. Michod.
     p.   cm.
 Includes bibliographical references and index.
 ISBN 0-201-40754-X
 1. Sex (Biology) 2. Evolution (Biology) I. Title.
QH481.M53 1994
575'.5—dc20
    94-13158
    CIP

Jacket design by Lynne Reed
Text design by Diane Levy
Set in 12-point Garamond 3 by Jackson Typesetting

12345678910-MA-97969594
First printing, November 1994

To Christina, Kristin, and Katherine

Contents

Preface

Sex is a problem. Indeed, sex is one of the great unsolved mysteries of life. The living world is sexual. Why? A bee lands on a blossom, a stag rears back his head in bellowing, a human couple lies exhausted in passionate embrace. The flower, the deer, the human, even the unseen virus—they all must have sex. But why?

The very question, Why is there sex? seems strange: sex is such a fundamental part of our nature—and, indeed, of much of life on earth. One immediate answer might be that sex feels good. But many sexual creatures have no brains—flowering plants, for example—and can hardly know what anything feels like. Another answer seems obvious: sex is for reproduction. After all, having babies is one thing most of us think about when we have sex—either we want them or we don't. But it is too simple to say that sex evolved just for reproduction, for nonsexual forms of reproduction exist: sponges, most forms of bacteria and viruses, and some kinds of lizards are a few examples.

Plato touched on the subject of why sex in *The Symposium,* where Aristophanes in a famous speech offered a solution in terms of repair—an idea that figures prominently in the story I have to tell. Plato's solution was embraced by Freud as it seemed to give biological content to Eros, the sexual instinct—the drive to live and preserve life. Freud discussed scientific theories of the origin of sex hoping to learn something about his postulated psychological instincts: the death instinct and Eros. Schopenhauer saw in sexual relations the source of all human conduct. In a passage from *The World as Will and Idea* he wrote:

> Eros is the first, the creator, the principle from which all things
> proceed. The relation of the sexes ... is really the invisible central
> point of all action and conduct, and peeps out everywhere in spite
> of all veils thrown over it. It is the cause of war and the end of
> peace; the basis of what is serious, and the aim of the jest; the
> inexhaustible source of wit, the key of all illusions and the meaning
> of all mysterious hints.

Undoubtedly, there are many other great thinkers who have involved
themselves with the problem of why sex. However, in the study of
life, all "why" questions ultimately find their way into the realm of
evolutionary biology—the science of how and why the living world
has come to be the way it is from its rather modest beginnings on
the primordial earth some 4,000 million years ago.

Charles Darwin mentioned sex, and its role in making variations,
in his great book *The Origin of Species*. He took it as a "law of nature"
that intercrossing was necessary for life because it created hybrid
vigor. In a later book devoted to plants, *The Effects of Cross and Self
Fertilisation in the Vegetable Kingdom* (1889, p. 462), he concluded
that "It has been shown in the present volume that the offspring
from the union of two distinct individuals, especially if their progeni-
tors have been subjected to very different conditions, have an immense advantage in height, weight, constitutional vigour, and
fertility over the self-fertilised offspring from one of the same parents.
And this fact is amply sufficient to account for the development of
the sexual elements, that is, for the genesis of the two sexes." He
was less than clear concerning the reasons why hybrid vigor was so
universal, probably because he lacked a correct theory of genetics.
Darwin's explanation is part of the story I have to tell, but it is not
the whole story, for it cannot account for the origin of recombination
and diploidy, two basic aspects of the sexual cycle. He knew of
examples of virgin birth, asexual reproduction, but felt that these
cases were poorly understood and all but dismissed them. He did
not seem to appreciate the real problem they posed. These glaring
exceptions to Darwin's "law" of intercrossing (sex) stand even today
as persistent reminders that all questions have not been answered
concerning the evolution of sex. That sex is widespread in the living

world is clear—but why should this be? If we take its preeminence in the biota as evidence of its advantage, how can asexual organisms like dandelions and some lizards do without it?

It was the great German biologist August Weismann who first defined the mortal and immortal in biological terms and who also, towards the end of the last century, first addressed the problem of sex in evolutionary terms. According to Weismann, sex accelerates evolution by providing a continual store of variability upon which natural selection can act. Weismann's voice on the matter is still heard today, enshrined as it is in textbooks on the subject. With the rediscovery of Mendel's laws at the turn of the century, the new science of genetics was born. Pioneers in modern genetics, specifically R. A. Fisher and H. J. Muller, recast Weismann's ideas in terms of the new genetics and the problem was solved. Sex accelerates evolution and that's a good thing. Case closed.

It wasn't until the 1970s that an American, George Williams, and an Englishman, John Maynard Smith, reopened the case for sex, but it turned out to be a case *against* sex. The first order of business was to lay out the costs of sex for all evolutionary biologists to see. And costly sex is.

Sex requires a huge commitment of time, energy and resources. Humans are intimately acquainted with the magnitude of the effort required; most are preoccupied, from adolescence, with finding suitable mates, mating, and keeping mates. Other sexual species have it no easier. Consider the peacock, which carries around a set of outlandishly ornamented tail feathers that serve to lure peahens but that also attract predators. Sexual reproduction also limits the number of genes a plant or animal can bequeath to its offspring. Evolutionary theory holds that the most fundamental aim of any living thing is to preserve its genes by passing them on. The asexual bacterium replicates all its genes every time it reproduces. However, a sexual creature contributes only half its genes to each offspring; the other half comes from its mate. These are the genetic costs of sex. There is also the cost sex levies on the reproductive potential of a species as a whole. Males make up roughly half of any population but, in the vast majority of species, contribute nothing save their genes to the next generation; females usually bear the entire responsibility of caring for offspring. (Nature's rare exceptions are some human fathers and the males of some species of birds, fish and insects.) In species that contain only asexual females, all individuals can produce and care for offspring, so the population has a greater reproductive potential by a factor of about two. There are still more costs. Obviously, sex requires intimate contact between two organisms. Infectious genetic elements have seized upon sex as an opportune moment to transfer themselves from one body to another and cause such calamities as AIDS. The costs of sex are huge and must be paid each and every time a sexual creature mates and reproduces.

Once the costs of sex were on the table, it became easier to appreciate the magnitude of the problem of sex and the nature of the task at hand for evolutionists. For any trait to evolve by Darwin's principles of natural selection, it must be beneficial to the genes involved. Yet, the sex trait appears to be more costly than beneficial. Weismann, Fisher and Muller had explained sex in terms of a benefit for the whole species—that sex speeds up its evolution. This would take a long time—many, many generations—to occur. In the meantime, there would be nothing to keep asexual females from spreading in the population and converting sexual populations into asexual populations. Asexual females, after all, would not have to pay the costs of sex, and so in the short term they would be at an advantage. The case against sex was even better than this. Maynard Smith and Williams showed that the benefit envisioned by Weismann and his disciples did not even exist in many situations.

This is where matters stood in the mid 1970s. For over 100 years, the problem of sex had appeared to be solved. All of a sudden, one of the world's most pervasive and costly adaptations had no explanation at all. You may not have noticed, but the resulting crisis was heard in laboratories and lecture halls around the world. A new generation of evolutionary biologists heard the battle cry and went on the offensive. Again, for the most part, they pursued Weismann's battle plan—sex creates variation and so sex must be beneficial in varying environments. Why, they asked, is sex beneficial in varying environments? This story is told in Chapter 5.

Twenty years of work have returned a simple answer. For sex to be advantageous in changing environments, it is not enough that the environment simply changes, or that it is unpredictable—it is not even enough that the environment be completely random. For sex to be advantageous, the association between two relevant states of the environment must continually flip-flop each and every generation. To put matters in the simplest terms possible—the world must be fickle.

But in many regards the world is predictable and not fickle. One day is like another. Offspring often find themselves with the same challenges as their parents faced. Some aspects of the world may be fickle, perhaps the world of rapidly evolving parasites is one, but for the most part the conditions needed by Weismann and his disciples are not general enough to explain sex. They are not up to the task at hand, because sex is found everywhere and in almost every situation.

Since the discovery of DNA almost forty years ago, the landscape of modern biology has changed drastically. However, the problem of why sex evolved, the queen of all problems in biology, has remained untouched by the DNA revolution, until recently. Remarkably, this new knowledge has taken us back to Plato's solution as told by Aristophanes. Sex keeps genes healthy.

Sexual systems are quite diverse at the level of behavior and morphology. However, the effect of sex at the level of DNA molecules is quite constant. Sex involves the breakage and rejoining of DNA molecules (usually) taken from two separate parents. Biologists term these characteristics *recombination* and *outcrossing*. Recombination plus outcrossing is what sex comes down to at the level of DNA. Why recombination and outcrossing? Because the DNA molecule gets broken and it must be put back together again.

The consequence of sex that figures prominently in my story is reflected in the contrast between a baby's youthfulness and its parent's maturity. Sex between aging parents produces babies that are youthful. The regenerative and rejuvenating powers of sex have appeared time and time again in psychology, philosophy, literature and art. Before Weismann, the view that sex rejuvenates life had been discussed in the scientific literature. This view soon fell into disfavor, because scientists could not explain how sexual reproduction worked any better than asexual reproduction in this regard. As a result, August Weismann was able to dismiss the idea by saying "twice nothing cannot make one"—meaning that there is no apparent advantage in using two aging cells, from two separate organisms, rather than a single aging cell, from a single asexual organism, to produce a new plant or animal.

In recent years, armed with an understanding of how DNA operates, Weismann has been proved wrong: twice nothing *can* make one. Specifically, two error-ridden DNA molecules, one from each parent, become whole, and thereby regain youthfulness, when passed through the sexual cycle. Coping with genetic errors is the primary biological consequence of sex. Sex keeps genes healthy. To make repairs, good spare parts must be at hand. The problem is how to keep backup copies of genes in good working order, in case they are needed for repair. As we will discover, this requires keeping tabs on mutations, which have a tendency to accumulate in cells. Sex helps to keep the numbers of mutations within manageable limits. Sex also helps avoid the potentially harmful effects of homozygosity, while keeping genes in good repair. In short, we will learn that sex evolved as an elaborate system for repairing genes, managing mutations, and avoiding homozygosity. The main theory is developed in Chapters 6 and 7 and reconsidered as a whole in Chapter 8. The nonspecialist will find these chapters challenging, but I hope that he or she will be rewarded for their effort.

But the story I have to tell is more than just why sex exists. It is also about why it matters that sex exists. A sexual world is quite different from what we can imagine an asexual world would be like. A sexual world is one large extended family, interconnected and interwoven like a rich tapestry. Yet within this tapestry we find the recognizable and distinct entities termed species. Why species exist is the question Darwin wanted, but was unable, to solve. Interconnectedness and distinctness are two features of life that are able to coexist in the same tapestry because of sex. (We consider this problem in Chapter 10.)

We are attracted to nature because of the marvelous adaptations we find there—beautiful ornamentations, intricate designs, expertly engineered structures. These features are all part of the mystery of life and the mystery of organisms. By understanding organisms, we hope to understand some part of ourselves. But, because of sex, organisms are of only fleeting existence, each born unique and each soon to die. Sex and death are uneasy partners in the process of life (Chapter 4). As we will find in Chapter 9, because of sex, organisms have no enduring value in the evolutionary process. What matters is for the process of life to continue to unfold, as it has done since its beginnings.

By managing the many errors and threats to DNA, sex has enabled life to diversify and expand and, hopefully, to continue to do so forever. Although every living thing is mortal, born to eventually die, life itself appears to be *immortal.* Since originating on this planet, life has continually been passed down from parent to offspring like a family heirloom and shows no signs of petering out. The continued life-giving powers of DNA reside in its well-being as a molecule and carrier of information. As we will discover, sex plays a key role in maintaining the well-being of DNA and, as a result, gives life immortality. As the prophet Diotima tells Socrates in Plato's *The Symposium*, "Its object is to procreate and bring forth in beauty . . . because procreation is the nearest thing to perpetuity and immortal-ity that a mortal being can attain." Humans do not remain forever the same, as do gods, she further explains, but through procreation "the losses caused by age are repaired."

Why Eros *and* Evolution? In early Greek mythology, for example in Hesiod's *Theogony* (c. 725 B.C.), Eros was one of the primeval forces of nature and emerged out of Chaos and Earth. Bulfinch (1971), describes a myth in which all begins with Eros.

> There is another cosmogony, or account of the creation, according to which Earth, Erebus, and Love were the first of beings. Love (Eros) issued from the egg of Night, which floated on Chaos. By his arrows and torch he pierced and vivified all things, producing life and joy.

In these myths, Eros is the embodiment of all harmony and creative power in the universe. In later mythology, Eros was a lesser god, the son of Aphrodite and Ares. His role was reduced to a more playful one as the constant attendant of his mother, the goddess of love. He became the mischievous perpetuator of love through darts of desire shot from his bow. Yet, he still maintained his youthfulness and creative power. His mother, Aphrodite, often complained that he stayed a child. In the famous myth of Eros and the mortal beauty Psyche, Psyche attains the immortality of gods by overcoming her many challenges and sufferings. According to Bulfinch (1971), the story of Eros and Psyche is an allegory in which Psyche refers to the human soul "which is purified by sufferings and misfortunes," by Eros and is thus prepared for immortal peace and happiness. In psychiatry, Eros refers to the instinct for self-preservation. In Freud (1961) the Eros becomes equivalent to the life instincts, the sexual instincts, which work "towards a prolongation of life" by seeking to "restore an earlier state of things."

Like Eros, sex is both a creative force and a conservative force. Eros seeks immortality by restoring a previous state of affairs in which the organism was whole. This theme can be found in the poems of antiquity and in the texts of psychoanalytic theory. It is most clearly expressed in Plato's *Symposium.* The theme of my book is how sex restores the DNA molecule to an undamaged state. Sex overcomes the many genetic errors—damage and mutations—that threaten life, and in so doing the DNA molecule becomes whole. Sex maintains the well-being of genes, and through their immortality, sex provides for the continuation and immortality of life.

I have written this book for two distinct audiences: the specialist working in biology and the interested nonspecialist curious about the mystery of sex. I have tried to address the needs and interests of both kinds of reader. For both readers, I hope to convey the mystery of sex and the challenge of fully understanding its significance. To be convinced by a hypothesis, the specialist requires a sound logical argument along with a thorough treatment of the data. I have focused on providing a clear argument, but I have been selective in choosing what facts to present. The nonspecialist should not be bothered with peripheral issues or with overly technical discussions. I have tried to keep to the main point in the text, while using the chapter notes for supporting material and references. All of the ideas in the book are based on work that has been published in the scientific literature. I would like to thank the National Science Foundation and the National Institute of Health for their support of my research (Grants NSF DEB 89-10191, NSF DEB 8118248, NSF BSR-8415436, NIH GM 36410, NIH HD19949). Many of the ideas were developed in collaboration with my colleagues and friends at the University of Arizona: Harris Bernstein, Henry Byerly and the late Fred Hopf. I am especially indebted to my friend and colleague Fred Hopf. I will never forget him, his sharp mind, and his enthusiasm for science. The many long hours we spent arguing and discussing evolution were some of my most memorable moments in science. I am grateful to my colleague Charles Sherry for suggesting, some years ago, Plato's *Symposium* to me. I am also grateful to Harris Bernstein, Henry Byerly, Alec Michod, and my editor Jack Repcheck, for reading and commenting on the complete manuscript. More than anything, I am grateful for the Eros in my life: my parents, and the love, support and inspiration of my wife, Christina, and daughters, Kristin, and Katherine.

Eros *and* Evolution

◆ 1

Why Sex?

*All men are brothers. we like to say. half-wishing some-
times in secret it were not true. But perhaps it is true.
And is the evolutionary line from protozoan to Spinoza
any less certain? That also may be true. We are obliged.
therefore, to spread the news, painful and bitter though
it may be for some to hear, that all living things on
earth are kindred.*

—Edward Abbey
Desert Solitaire

Before I noticed it, it was gone. Just a momentary streak of light
in the night sky. As it came to the end of its journey, the meteor
quietly disappeared with a flicker, or so it appeared to me as I
witnessed the event from many thousands of miles away. How long
had it been up there? How far had it journeyed, free and unencum-
bered in the vacuum of space? What worlds had it passed before
entering mine? The interaction started from afar. Only an impercep-
tible nudge at first, earth's gravity gently cajoling the unsuspecting
rock off its course into its fated mission. Three bodies interacting,
a rock, the planet earth, and me. If I had blinked, I would have
missed it. Our paths crossed for just a few seconds that lonely sum-
mer night, as I looked up into the vast desert sky.

Forces are at work, keeping the world together by attracting one

1

thing to another. In the physical world, the basic forces—gravity, the electromagnetic force, weak and strong nuclear forces—have been identified, although they are not fully understood. But what about the living world: why are living things attracted to one another? A male praying mantis follows his mate dutifully, although after they have sex she may eat him. A robin perches in the shadow of a tree while her mate sings. A green turtle leaves its wintering ground on the coast of South America and undertakes a 3,000-mile journey to tiny Ascension Island in the middle of the Atlantic Ocean to mate. Why do they bother? We humans, why do we love?

Why Sex?

For the simplest virus, as well as for the sophisticated human, sex requires a partner. Partners must be attracted to each other. And they must have sex (as we will see). But what exactly is sex?

When we think of sex, we usually think of having sex. This is only natural, given the pleasure we derive from it. When we think of the problem of sex, our attention may turn to our own personal inadequacies or sexual hang-ups. But there is a more universal problem of sex that is much broader and more basic than our feelings. It is the unresolved question of why sex exists at all. What are the consequences of sex that make it so important to us and so widespread in nature? The answer to this question lies not in our own attitudes and feelings about sex, but deep in our evolutionary past.

A simple, if incomplete, answer is that sex is for reproduction. After all, having babies is one thing most of us think about when we have sex, because we either want them or we don't. However, sex is not always associated with having offspring. Single-celled organisms, such as bacteria and viruses, have sex without reproducing (I'll explain how they do this in Chapter 6). And they're surely not in it for pleasure, because these creatures have no feelings at all. Such simple organisms, which are most similar to the kinds of things that lived when sex first evolved, must be having sex only for the most fundamental biological reasons. But what are these reasons?

It may come as a surprise to you, but biologists disagree strongly concerning the purpose of sex. Why sex exists at all is one of the great mysteries in biology. There is, however, one thing biologists can agree on—the costs of sex are plentiful. Consequently, the evolution of sex has proved difficult to explain, since at first glance the costs are large and easily appreciated while the benefits seem unclear.

Costs of Sex

First, there are genetic costs. Evolution favors traits that effectively transfer their underlying genes to future generations. Traits such as eyes for seeing, hands for grasping and legs for running are obviously beneficial and help the organism pass on its genes (including the genes for these traits) to offspring. Yet the sex trait *seems* to do exactly the opposite. Each sexually reproducing parent passes on only one-half of its genes to its offspring. The other half of its offspring's genes comes from its mate. An asexual parent produces offspring by itself. For example, a mother whiptail lizard, who lives in the deserts of the American Southwest, passes on all of her genes to her offspring, because she is able to reproduce by virgin birth, without male sperm.

So, a sexual parent passes on only half as many of its genes as an asexual parent, all other things being equal. A difference of 50% in the propagation of genes is a huge difference, especially when compared to most other traits that vary in a population. The differences in beak size between birds in a forest make at most a 10% difference in the birds' ability to have offspring and pass on genes. Even small differences between traits, on the order of 1%, in their ability to pass on genes have significant effects when accumulated over many thousands of generations. An initial twofold difference between a sexual and an asexual parent in the propagation of genes is so large as to be unimaginable!

Secondly, sex is costly in terms of the energy, time and resources required to find a suitable mate and in terms of mating itself. Humans are intimately acquainted with the magnitude of the effort required; most of us are preoccupied from adolescence with finding a suitable mate, and much of our adult life is devoted to finding or keeping one. And we're not alone—the rest of the living world is likewise obsessed. Consider the outlandish ornamentation of the peacock's tail (which attracts predators as well as peahens), the humorous displays of male sage grouse (which take time and energy away from feeding), and the burdensome antlers of male deer (designed to appeal to females, or to win them in battle with other males). What's more, in many organisms mating requires intimate contact between the two parents, which provides an opportunity for parasites, viruses and bacteria to move from one body to another—and cause such problems as AIDS, herpes, and chlamydia. Indeed, some biologists believe that sex was invented for the benefit of such infectious agents—so that they could move from one body to another—and not for the benefit of the parents at all!

Finally, there is the cost of producing males. Males make up roughly half of any sexual population but, in the vast majority of species, they contribute nothing save their genes to the next generation. Most males depart after mating (usually to mate again) leaving the mother with the entire responsibility of caring for the children. Some human fathers, and the males of some species of birds, fish, and insects, for example, stay around to provide time and resources to care for their young. But these are nature's rare exceptions. Most males do nothing to care and provide for offspring. In species composed entirely of asexual females, in contrast, all individuals can produce and care for offspring, so an asexual population should be able to produce about twice as many offspring as a sexual population.

The Paradox of Sex

The more costly a trait, the greater its benefit, so that its net effect on the fitness of the organism is positive. Otherwise, the trait could not evolve. In the case of sex, the benefits are not as obvious as the costs. This is the paradox of sex. It seems on the face of it that sex should *not* have evolved. Yet, not only has sex evolved, but it has spread throughout the living world.

There are two main theories concerning the benefits of sex. The standard view (which will be the subject of Chapter 5) is that sex allows for the production of more diverse offspring than could be produced asexually. (A sponge that makes an exact copy of itself produces nothing essentially novel. But a beagle that mates with a Pekinese can produce a puppy that looks like none of its ancestors.) According to the standard theory, this ability to produce genetic variation in turn enhances a species' ability to adapt to new and complex changes in the environment. A new twist to this view argues that variation helps long-lived organisms cope with parasites that have much shorter life spans and so can evolve quickly.

According to this "variation view," sex does not benefit the sexual individuals themselves, only their offspring. In other words, if we consider two organisms differing *only* in that one is sexual and the other not, the sexual one does not survive to reproduce any better than the asexual one. In fact, given the costs of finding a mate and otherwise being sexual, a sexual parent may actually be worse off in terms of surviving and reproducing than an asexual one. However, by having a diverse group of offspring, a sexual parent may end up with more surviving offspring than an asexual parent. If, for example, the habitat becomes colder, only offspring with heavy fur coats may be able to survive. Even if this type did not exist in the population, two sexual parents may have a chance of producing it through chance recombinations of their genes.

The eminent German biologist August Weismann first argued in favor of the variation view toward the end of the last century. Today, it is the dominant theory for the evolution of sex, presented as dogma in most textbooks. As we will see, however, this view has several serious flaws. Chief among them is the fact that sex undoes what it creates. Sex may produce a beneficial new combination of genes in one generation only to break it apart the next. Consider, for instance, a pair of mating birds in a forest in which the foliage has changed color slightly. One of their offspring, by virtue of a chance recombination of genes, might have feathers that blend in particularly well with the new flora and thus offer better camouflage. But when this bird matures and reproduces, mixing its traits (and genes) with those of its mate, the combination of genes that produced its coloring is likely to be destroyed, turning the feathers of its offspring back to the original shade or to another color entirely. Only rarely does sex produce something beneficial, but when it does the beneficial combination is soon destroyed by the very process that created it. I believe this view is untenable as a general theory of sex, although it may apply to special circumstances, as discussed in Chapter 5.

The second explanation (this is the subject of Chapters 6, 7 and 8) is based on the role that sex plays in coping with genetic error. Parents pass on their genes by producing offspring. Offspring produced sexually are more fit, because their DNA is in better condition than the DNA of offspring produced asexually. However, sex is not really for offspring, for parents, or even for organisms for that matter, even though having sex is pleasurable to them. The most basic biological consequence of sex is not even reproduction, but rather the health and preservation of the genes, or DNA molecules, carried by the organisms that practice sex. Sexual creatures vary widely in size, shape and behavior, yet the cleansing and rejuvenating effects of sex at the level of the DNA molecule are always the same. Sex is for genes.

Complementarity and Attraction

Deoxyribonucleic acid, DNA, is designed to store and transfer the information needed to make and maintain a living thing. Like all systems capable of storing and transferring information, DNA uses an alphabet to encode the data. The English language, for example, uses the set of 26 Roman characters. DNA's alphabet, on the other hand, contains only four letters, A, T, G, and C, which stand for four types of chemical compounds called nucleotide bases: adenine, thymine, guanine and cytosine. DNA comes in strands, which consist of a sequence of these bases, and their order has meaning, just as the order of letters in a word gives it meaning. The linear sequence of the four nucleotides in a single set of plant or animal genes, which may consist of as many as a billion nucleotides, contains all the information necessary to make that organism. Each DNA molecule, sometimes called a chromosome, is punctuated into many shorter segments of approximately 1,000 to 5,000 nucleotides each. These shorter sequences are called genes. If you think of all the genes needed to make the organism as a book of instructions, then a gene, in turn, may be likened to a single sentence in that book. Individual genes determine specific features that make up the organism. For example, different genes determine different eye color, hair color, height, skin texture and stature. Virtually every characteristic a living creature possesses—from size and shape to sound of voice and behavior—is to some extent affected by the genes it carries.

Each DNA molecule contains two strings of nucleotides, or strands, that are complementary to one another (see figures in Chapter 2). Both strands contain the same information, but one strand is the complement of the other, much like positive and negative images in photography. These strands are held together and attracted to one another by weak physical forces, called hydrogen bonds, which operate between complementary nucleotides. For physical and chemical reasons, *A* is attracted to *T*, and *G* is attracted to *C*. This is to say *AT* and *GC* combinations are more stable than, say, *AC* and *GT* combinations. Thus, the two single strands in one DNA molecule have complementary sequences of nucleotides. If the sequence in one strand is *ATTCGG*, then the sequence in the other strand must be *TAAGCC*.

Sex is Recombination with Outcrossing

Human sexual reproduction begins with the making of germ cells (eggs in the case of females and sperms in the case of males). Meiosis is a special cell-making process that occurs in the gonads by which we (and other complex creatures) make germ cells. During meiosis, the DNA molecules we inherited (half of them from our father and half from our mother) are broken into pieces and reconstructed in new configurations. The new DNA molecules, each one unique, are packaged in eggs or sperms and passed on to our offspring.

There are two aspects of human sexual reproduction that are universally found in all sexual creatures: recombination and outcrossing. *Recombination* refers to the physical breakage and rejoining of DNA molecules. *Outcrossing* refers to the fact that the DNA molecules involved in recombination come from two different individuals from the previous generation: our mother and father.

Recombination is a general feature of DNA and need not involve outcrossing. Agents which damage DNA can increase the occurrence of recombination whether the cells are reproducing sexually or not. For example, sunlight increases the occurrence of recombination in yeast cells that reproduce by an asexual process called mitosis. Although recombination can occur during mitosis, it is especially frequent during meiosis and other sexual processes.

The Theme of the Book—Coping with Genetic Error

Recombination repairs damaged genes—whether in mitotically produced asexual cells or in meiotically produced germ cells. It is directly beneficial to the DNA molecule, and hence to all cells and organisms, both parents and offspring. But recombination itself is not sex. Why did recombination become associated with outcrossing during sex?

In Chapter 2, we consider how sex likely originated in simple unicellular creatures such as bacteria and viruses. Most of the time, these cells contain only one copy of each gene (they are said to be *haploid*); outcrossing is needed to bring spare genes into a haploid cell to repair gene *damage*—detectable errors in DNA. However, in multicellular organisms like humans, almost every cell contains two copies of each gene (they are said to be *diploid*); so there should be spare genes available in the cell for gene repair. Outcrossing is no longer needed in these organisms as a source of spare gene parts. So what is the purpose of outcrossing in complicated diploid organisms? As we will discover in Chapters 7 and 8, sex in diploids allows for gene repair without the harmful consequences of homozygosity.

Damage is detectable in the DNA molecule and for this reason can be repaired by recombination. A second kind of error, termed a *mutation*, also occurs in the DNA molecule. Unlike damage, mutations cannot be detected at the level of the DNA molecule. Because they go undetected, recessive mutations accumulate in diploid cells and destroy the information contained in the spare gene parts. (Have you ever reached for a backup copy of a computer file and found that it, too, was damaged?) We will learn in Chapters 5 and 7 that outcrossing sex helps keep mutations from accumulating, so that both copies of each gene are in good shape if one happens to be needed for gene repair. This, in a nutshell, is my theory about why sex evolved. The remainder of the book is devoted to exploring why it matters that sex evolved.

Sexual World

The two components of sex, recombination and outcrossing, may not appear very interesting at first glance—not very "sexy." But they are the features held in common by all sexual organisms, whether the simple virus or the more sophisticated human.

As we will discover in the pages that follow, recombination and outcrossing, the breakage and rejoining of DNA molecules from different parents, have profoundly affected the nature of the living world. A sexual world is made up of species that are distinct and recognizable. Yet these species are interconnected. In a real informational sense, the sexual world is one. A sexual world is immortal, although it contains organisms that are mortal and of no lasting significance.

Organisms are fascinating to us because of their wondrous adaptations—their inspiring designs and intriguing behaviors. We wish to understand organisms, in part, to understand ourselves. But, to understand ourselves we must go beyond organisms—and understand their genes and the genes' incessant capacity to recombine. Because of sex, organisms are of only fleeting existence, each born genetically unique and each soon to die. An organism has no enduring value in the evolutionary process. Sex does not exist for organisms. Sex was created for genes and genes are forever.

Survival of the Fittest

Even Darwin's notorious phrase "survival of the fittest" is best understood as referring, not to organisms, but to gene lineages. Darwin did not use the phrase "survival of the fittest" in the first edition of *The Origin of Species*, which was published in 1859. By the sixth edition, published in 1866, at the encouragement of his supporter Thomas H. Wallace, Darwin embraced Herbert Spencer's now notorious phrase. Natural selection and survival of the fittest had become synonymous, and Chapter IV of the great book was retitled "Natural Selection or the Survival of the Fittest." Darwin accepted Spencer's phrase because it succinctly summarized what he expected to be the outcome of natural selection: fitter organisms surviving and predominating over time.

The phrase "survival of the fittest" has been seized upon by critics of the theory of evolution and accused of being nothing more than an empty tautology. Who are the fittest? Those organisms that survive. If "survival" is the very criterion of "fitness," the principle of natural selection, as asserting "those organisms that survive are those that survive" is truly circular. So the tautology challenge runs, to the exasperation of evolutionary biologists who have little fear that their science can be reduced to empty truisms.

A definitive way to diffuse the charge that "survival of the fittest" is tautological, is to show clear cases where it is false. We will find many such examples in the pages that follow. These examples all stem from sex and the way it affects the dynamics of natural selection. So the phrase "survival of the fittest" cannot express a tautology, since tautologies cannot be false. More importantly, we will learn from these examples that fitness is not a property of organisms, in spite of what Darwin and Spencer said, at least not if "fitness" is to be used in a predictable and nontautological way. And all of this because of sex.

Immortality

Although every living thing is mortal, born to eventually die, life itself appears to be *immortal*. It has been around for some 4,000 million years now and shows no signs of petering out. Life doesn't begin at some specific moment each generation (as a legal reading of the abortion/choice debate might lead you to believe). Rather, it originated once, and has since been passed on from parent to offspring throughout the eons. It is possible, in principle, for each of us to trace a continuous line of ancestors back to the first living thing. Life's immortality resides in the well-being of the genetic material, DNA. And, as we will see, there is good evidence that sex plays a key role in maintaining the well-being of DNA and hence the immortality of life.

This role is reflected in the way sex appears to rejuvenate life; it allows two aging adults to produce a youthful baby. It was this obvious fact that led nineteenth-century scientists to assume that sex exists entirely to rejuvenate life. Their view fell into disfavor, though, because they could not explain how sexual reproduction worked any better than asexual reproduction in this regard. Weismann had dismissed the idea with the statement "twice nothing cannot make one"— meaning there is no apparent advantage in using two aging cells, from two separate organisms, rather than one aging cell, from a single asexual organism, to produce a new plant or animal. In recent years, however, armed with an understanding of how DNA operates, Weismann has been proved wrong: twice nothing *can* make one.

Distinctness and Species

In 1859, Charles Darwin asked the following question, which he set out to answer in his revolutionary book, *The Origin of Species*:

> Why is not all nature in confusion instead of the species being, as we see, well defined? . . . Why are not all organic beings blended together in an inextricable chaos?

With this question, Darwin raised one of the most difficult and still-unresolved problems in biology. Why is nature organized into the recognizable and distinct groupings we call species? It was this question that Darwin wanted to solve in his book, as its title advertised. *The Origin of Species* revolutionized our views of the world and brought evolutionary thinking into the mainstream. In this monumental work, Darwin showed considerable insight and clarity on issues that only recently have been confirmed. In many details of profound significance he was right. However, with regard to the origin of species, the central problem of the book, Darwin was wrong.

There are two characteristics of the living world that Darwin wanted to explain: the existence of species and design. Good design is apparent in the perfection with which living creatures are adapted to their environment: the intricacy of the human eye, the power of a lion's paw, the match of the pollinator to its flower. Before Darwin, the existence of well-designed organisms implied for most people the existence of a divine being, a creator, just as a watch implies a watchmaker. Darwin destroyed this logic by showing that it was not a divine being that designed the living world, but a natural process—natural selection. Natural selection is a process of profound significance in the world. We know of no other process—in the physical or living world—that can create order out of disorder (in systems that are maintained far from thermodynamic equilibrium). Even physical systems, such as lasers and the convection patterns in heated liquids, obey nonlinear dynamics similar to those implied by Darwin's principles.

The second problem considered by Darwin was the existence of species. Cultures without the formal study of modern biology typically recognize the same natural categories, or species, as a professional biologist. Different species are recognizable because they are distinct—there are gaps between them. Why are there holes in the tapestry of life? Why is nature not in confusion, as Darwin asked? Darwin dearly wanted answers to these questions, but, as we will see, Darwin's answers simply don't work. The answer to this problem, as with so many fundamental problems in life, stems from sex.

Sex requires two—and the two must find one another. As already mentioned, all sexual creatures expend considerable resources finding and keeping mates. The less numerous the species, the more effort is required to find a suitable mate. For example, a certain flowering plant expends more effort in nectar production when rare than when common, since it has more difficulty attracting pollinators when it is rare. For these reasons, sex has an intrinsic cost of rarity. As we will discover, this cost of rarity profoundly affects the character of evolution in sexual populations and provides an answer to the question of why species exist.

Unity from Connectedness

A child possesses some of the same features her mother and father have, as well as new ones not present in either parent. Our new baby has her mom's cheeks and my eyes, but where did she get her wonderful golden hair? Sex mixes and combines the genetic information from two individuals. This leads to yet another basic characteristic of life that stems from sex—unity through connectedness.

Because each of us has two parents, and each of our parents also have two parents, we have four (2×2) grandparents. After three generations we can count 8 (2×2×2) ancestors. In the diagram below we go back 5 generations to 32 ancestors. What if we were to go back 100 generations (a small amount of time when compared to the whole history of life)? We would be able to count some 2^{100} ancestors, which equals approximately 10^{30}, or 1,000,000,000,000,000,000,000,000,000,000 individuals. This calculation is true for each and every sexual creature. It is impossible to enumerate all of the organisms alive today. For just one species of bacteria, there are probably over 10^{20} living cells. All told, in 100 generations, that would mean there must have been more ancestors than atoms in the universe. Of course, this is far more organisms than could ever have lived. Clearly something has gone wrong in our calculations.

Our error lies in the assumption that all these ancestors were different individuals. In fact, we share ancestors both with each other and with all other living things. In Figure 1, we show the ancestors for a single sexual individual going back in time just four generations. In just four generations, space becomes limiting in the diagram and the branching lines start overlapping. We can think of this overlapping of lines as representing the overlapping of ancestors. In a finite world (like the diagram), some ancestors must be the same individual at some point in the past. No matter how small we drew each branching line in the diagram, the lines would start overlapping at some point. Because of sex, we simply must share ancestors with

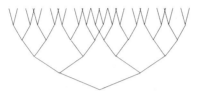

Figure 1. Ancestors

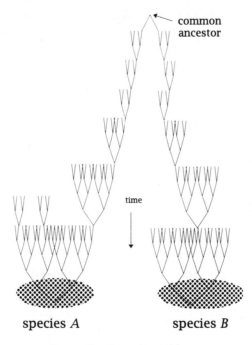

Figure 2. Sexual world

each other and with all living things. And because we share ancestors, Edward Abbey is correct—"all living things on earth are kindred."

In Figure 2, we see a sexual world, one with interconnectedness, distinctness and immortality. Two arbitrary species are shown, distinct from one another in the present, yet ultimately connected through a common ancestor, perhaps far in the distant past. Only one such ancestor is shown, for just two individuals in only two species. Even for these two individuals, there were other common ancestors, and, of course, there are many, many other individuals in these and other species. A sexual biology gives a pattern of distinctness at each moment in time cast upon a web of interconnectedness from the past. This web continues deep into the past and on into the future, hopefully forever, so long as we humans don't interfere.

As you scan the pool of unfamiliar faces next time you ride the subway or rush through a busy airport, pause for a moment. From the blur, pick out a single face and look into it. Look hard. And remember there was a person somewhere, sometime, not too far in the distant past, that connects you to that person.

But do not limit your attention just to human ancestors. Take a walk in the woods or go to the zoo. Pick a creature, any creature. An ape, a tree, a fish, a worm. For each creature, there exists an ancestor or relative that the two of you share. A continuous line of ancestors can be traced backward in time to that shared ancestor and then forward in time through its descendants to you. Your most recent relative to the ape was an apelike creature that lived just 5 million years ago. For the tree, your most recent relative lived some 800 million years ago—a single-celled creature similar to a modern-day paramecium. For the fish, your most recent relative was a fish that lived in the oceans about 370 million years ago, before living things invaded land. For a worm, your most recent relative was an ancient worm-like creature that lived in the oceans some 600 million years ago. There is one ancestor that all past, present and future creatures share—the first living thing. By considering the first living thing, how it arose, and what its problems were, we will discover how sex likely originated.

The Early Replicators

Was there to be any end to the gradual improvement in the techniques and artifices used by the replicators to ensure their own continuation in the world? There would be plenty of time for improvement. What weird engines of self-preservation would the millennia bring forth? Four thousand million years on, what was to be the fate of the ancient replicators? They did not die out, for they are past masters of the survival arts. But do not look for them floating loose in the sea; they gave up that cavalier freedom long ago. . . . They are in you and in me; they created us, body and mind; and their preservation is the ultimate rationale for our existence. They have come a long way, those replicators. Now they go by the name of genes, and we are their survival machines.

—Richard Dawkins
The Selfish Gene

Imagine sitting cross-legged by the primordial pool on the primitive earth around 4,000 million years ago, leaning over the slime and goo and, aided with some magical magnifying glass, watching the very first replication of *THE* ancestral molecule. Tentative and awkward at first, making many mistakes. Nevertheless, it was a start. Imagine all that this molecule would come to be.

The Primordial Pool

Like a human fetus that blossoms from a single cell into a life filled with creativity, achievement and passion, from such modest beginnings, the molecule races forward in our dream, multiplying and diversifying, propelled by nothing but its own replication, guided only by its environment, into some 1,000,000 species and countless individuals, some now living, most long dead. But we have gotten ahead of our story, let us return to the molecule—or to be more exact, the molecular replicator.

The Molecular Replicator

A molecular replicator is a molecule that makes copies of itself. DNA and RNA are the best-known examples of such molecules. They are made up of long strings of nucleotide bases, which, as we know, behave like characters in an alphabet. Information in the first replicator was probably likewise contained in a linear sequence of characters.

But where did the characters come from and how were they strung together? We can only guess. One view is that the characters—the nucleotides—were available in a kind of chemical alphabet soup present in small pools or water droplets, created from physical and chemical reactions that occurred on the primitive earth. (In a famous experiment, scientists showed that by passing lightning-like electrical discharges through a flask containing nothing but methane, ammonia, water and hydrogen—thought to be present on the primitive earth—nucleotides can be produced.) At some point, these characters linked with one another to form short strings.

Recall that individual nucleotides adenine (*A*), thymine (*T*), guanine (*G*) and cytosine (*C*) are attracted to one another. In the case of DNA, *A* and *T* pair with each other, as do *G* and *C*. In the primordial pool, likewise, the elements of a string such as *AATG* (the "parent" strand) could attract the complementary characters available in the alphabet soup. These may then have linked up to each other to form a complementary string *TTAC* (the "daughter"

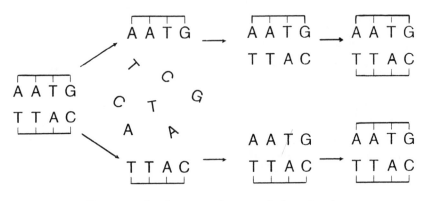

Figure 1. Replication of primordial molecules

strand). Such a pair of strands, held together by hydrogen bonds, is normally stable. However, under certain conditions (high temperature, for example) the bonds would weaken and the strands separate. The two single strands could then drift apart, and when conditions changed to again favor attraction, they would be free to attract new complementary strands and replicate as shown in Figure 1.

The forces of attraction and repulsion took the nucleotides, which are just physical chemical entities, and led them across life's doorstep into the world of the living—attraction and repulsion those two universals found in all interactions. Consider Freud's comments in a letter to Albert Einstein in 1932: "We assume that human instincts are of two kinds: those that conserve and unify, which we call 'erotic' (in the meaning Plato gives to Eros in his *Symposium*), or else 'sexual' (explicitly extending the popular connotation of 'sex'); and, secondly, the instincts to destroy and kill, which we assimilate as the aggressive or destructive instincts. These are, as you perceive, the well-known opposites, Love and Hate. Transformed into theoretical entities, they are, perhaps, another aspect of those eternal polarities, attraction and repulsion, which fall within your province." We will have much more to say about the Eros and Plato's *Symposium* in later chapters.

The cycle of strand building and separation, attraction and repulsion, shown in Figure 1 could continue indefinitely, given an ample supply of chemical building blocks, and thus the number of *AATG* or *TTAC* strings in the primordial pool could multiply. Since the parent strand serves as a template for the creation of the daughter strand, this process is known as template-mediated replication. Although there are considerable practical problems, one can imagine how, in principle, template-mediated replication started.

The capacity to replicate is a basic feature of life. A second basic feature is feeding. "Metabolism" is the more correct word and means the conversion of high-energy compounds (food) to low-energy compounds (waste). Life is order, and it takes energy to maintain that order. Highly ordered structures, such as organisms or molecular replicators, maintain their order by feeding and in the process the food becomes waste. For the molecular replicator, its "food" was simply the nucleotide building blocks, which must be in an energized state to link up with the next nucleotide in the string. Once the linking bond is formed, the nucleotides are in a lower energy state. Energy sources present in the environment, such as sunlight, may have energized the nucleotide building blocks in the first place.

What kinds of properties would these strings of nucleotide characters have? In other words, what would they look like and what would they do, besides copy themselves? Probably each string folded and twisted itself into a unique shape, determined by its particular linear sequence of characters and the physical properties of the environment, such as the acidity, salinity and temperature of the pool.

For these early molecules, shape was probably very important. Death of the molecule would result from breakage of the bonds that held the string together or changes to the characters themselves. If a molecule were especially twisted and compact, it might stand less chance of being broken apart. On the other hand, compact shape probably would not be conducive to replication. Because replication requires that nucleotides be able to move into positions next to their complementary characters in the replicating molecule, a more open shape would probably facilitate the copying process, while a more compact shape might retard it. In any event, the molecule's three-dimensional shape would strongly affect the molecule's capacity to survive and reproduce.

Evolution of Molecular Replicators

As the molecules copied themselves, mistakes probably arose quite often. Why? The physical forces that attract *A* to *T* or *G* to *C* are not completely faithful. Laboratory experiments with simple molecular replicators indicate that one out of every ten nucleotides copied is erroneous if replication occurs without the aid of enzymes or other cellular machinery (as it must have on the primitive earth). In comparison, the error rate for DNA copying in a modern cell is 1 mistake out of 10 to 100 billion nucleotides copied. So, for every ten characters in our early replicator, one "wrong" nucleotide would be placed in the daughter strand. (There might be a *C* across from an *A*, instead of the thermodynamically preferred *T*.) What happened then?

Because of complementary nucleotide pairing, mistakes can easily be copied. So the error—technically called a *mutation*—would be copied, along with the rest of the "correct" nucleotides, so long as the mutation did not interfere with the new molecule's capacity (based on its shape) to replicate and persist in its environment.

Once such a mistake copied itself, there would have been two different types of molecules in the population, one defined by the sequence *AATG* (or *TTAC*) and the other by the sequence *TCAC* (or *AGTG*). To the extent that these two molecules had different structures and properties, they may also have had different abilities to replicate or survive. The new molecule may have been more stable or able to replicate faster than the original one because of its particular shape. If so, it would have come to predominate in the population, while the old one might have gone extinct. Of course, it could

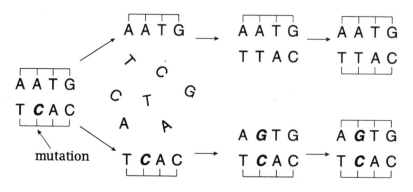

Figure 2. Replication of a mistake, a mutation

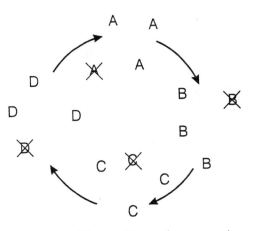

Figure 3. The world was born sexual

just as easily have been the case that the old type would have been better able to replicate or persist. Either way, the molecules would have gone through the process Darwin called natural selection, and the types with greater ability to survive and reproduce would have come to dominate the population.

In the Beginning, there was Sex

Individual replicating molecules were probably not isolated from one another. Parts of one molecule could easily break off and attach to other molecules. This continual breakage and rejoining of nucleotide strings is directly analogous to the recombination of DNA molecules that goes on during sex. In other words, the living world was born sexual. There was little individuality and much "promiscuity" at this early stage of life. It wasn't that evolution wanted it that way, it just so happened that nothing could be done about it, until, as we will see, the replicators became more advanced and improved their individual integrity.

The free exchange of parts (nucleotide strings) would have meant the creation of new sequences of nucleotides with new properties, some of which might have improved a molecule's capacity to survive and reproduce. What's more, the exchange of parts would have allowed for the repair of certain damaging errors that would often have occurred under the conditions of intense sunlight and heat prevailing on the primordial earth. Unlike mutations, damaging errors cannot be replicated and must be repaired for the DNA strings to continue replicating. Let me explain.

What kind of damaging error are we talking about? The ultraviolet (UV) light that makes up sunlight is harmful to nucleotide strings, even when they occur (as they do today) in DNA protected in the nucleus of a cell. So for naked molecules in the primordial soup, UV light would have been extremely harmful, especially when one considers the high intensity of UV light on the primitive earth (there would have been no ozone layer to help absorb the UV light). UV light can break nucleotide strings, permanently link one complementary strand to the other and thereby prevent it from separating and replicating, or cause spontaneous changes in the chemical structure of the nucleotide bases so that they lose their attraction for their complement. Under such conditions, some molecules might take on some very odd members—not characters in the alphabet (A, T, G or C) but things altogether different, say a ✿ or a ✍. This kind of error—technically called *damage*—might make it impossible for the molecule to reproduce at all.

Free exchange of replicator parts could help maintain a healthy population of replicators. I remember a friend in college who collected beat-up and run-down Volkswagens. He always was able to keep a few cars running by constantly incorporating good parts from damaged and otherwise inoperable cars. So long as all the cars in his junkyard were not defective in the needed part, he was able to keep some cars in good repair. As shown in Figure 4, a similar kind of repair can operate between damaged DNA molecules. If two damaged molecules have their errors at different positions in the nucleotide sequence they might both break apart (just by chance) and exchange halves, so that there would be two new molecules— one fully repaired and the other with both errors.

The newly repaired molecule could now go on to replicate and make fresh copies of itself, while the doubly damaged molecule would decay. This kind of selection sets biological repair apart from

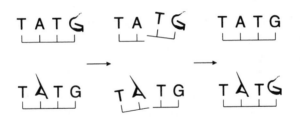

Figure 4. Repair of damage by recombination

the mechanical repair my friend practiced with Volkswagens. My friend was always looking for new junk cars. Starters would often break and soon the supply of working starters in his junkyard would be gone. His spare part inventory was usually precariously close to zero for some critical part. This is because Volkswagens, even ones that run, are not replicators—they do not make copies of themselves.

Recombination between replicators could maintain a relatively healthy population. The process of repair diagrammed above would also have helped the replicators get rid of mutations (improper nucleotide sequences) as well as damage. Free exchange would continually purge the population of errors, damage or mutations, by collecting them in wastebasket molecules. At the same time, recombination would generate error-free molecules that could continue replicating, maintaining a never-ending supply of spare parts and thereby giving the replicators a kind of immortality.

The First Cell

Template-mediated replication had its limits. The assembly of any structure is easier if certain tools are available. For example, a model airplane is easier to put together on a level table with good lighting and clamps and vises to hold the parts in place while the glue dries. Likewise, one can imagine how template-mediated replication might proceed faster and more accurately if there were tools to hold the strands in place, position the nucleotides, keep out unwanted dirt or harmful pollutants, and so on. Initially, physical objects available in a molecule's environment, perhaps clumps of clay or the surfaces of water droplets, might have been used to aid in the assembly of the daughter strand by stabilizing the intermediate structures. However, it would have been more advantageous for the molecule to be able to create its own tools—especially if this ability could be passed on to offspring to ensure their survival and replication.

Today, in living cells, DNA replication occurs within an environment similar to a factory of complicated machines and tools that aid the template-mediated process shown in the preceding figures. These tools are made of proteins. Like DNA, proteins are long sequences of characters, in this case consisting of amino acids rather than nucleotides. Because there are twenty amino acids, instead of only four nucleotides, proteins come in far greater varieties and so can exist in many different shapes. How and why did evolution produce the first protein?

Proteins owe their existence to DNA; the sequence of a protein's amino acids is encoded in the sequence of nucleotides in a particular stretch of DNA. We can only guess at *how* this translation of DNA information into protein was first accomplished. Ever since the discovery in the mid 1950s of the triplet code that translates genetic information into proteins (one amino acid for every three nucleotides), geneticists have speculated on the origin of the code and the first protein-making systems. We will not discuss this matter here. More relevant to our concerns is *why* this evolutionary advance first occurred. Fortunately, we have a better grasp of this question.

With their enhanced capacity to produce different shapes, proteins are more suited to make the tools that would help the replicators survive and reproduce. Furthermore, before it began making proteins, the nucleotide sequence itself had to accomplish all the tasks related to its survival and reproduction. We noted that the unaided DNA molecule encountered conflicts in what shape was best for it to attain. The optimal shape for replicating (an open shape with the nucleotides exposed) was more susceptible to death. But the optimal shape for avoiding death (a closed, compact shape) was difficult to replicate. By making tools specialized for each task (for example, one protein for protecting the sequence from death and one to help it replicate), the nucleotide sequence was better able to accomplish both tasks at once. By farming out these tasks to the proteins, the nucleotide sequence could specialize in a third task, storing (and acquiring through mutation and recombination) the information needed to make more proteins. This division of labor between the executive function (the proteins) and the legislative or instructive function (the nucleotide sequence) is familiar to us, as it exists in many social organizations and systems of government.

In biology, these two different responsibilities go by the technical names *genotype*, for the legislative/instructive responsibility, and *phenotype*, for the executive responsibility. Until the replicators evolved the capacity to make proteins, the genotype (genetic information) and the phenotype (structure or shape) were inextricably linked, because the same molecule had to do both. With the evolution of a means to translate the information stored in the nucleotide sequence into proteins, the genotype and phenotype could comfortably coexist without interfering with each other's function.

Useful tools, such as proteins, are likely made only at some cost to the replicator who happens to acquire, either by mutation or recombination, the ability to make them. After all, it takes time and energy to make the proteins, time and energy that could be spent replicating. So long as the benefits of the protein outweighed these costs, protein-producing replicators would be favored by natural selection. However, once protein producers evolved and predominated in the population, it would be advantageous for an individual replicator to stop making its protein and simply use the proteins made by other neighboring replicators. Such a "selfish" replicator could avoid the costs of protein production but still reap the benefits of the proteins made by its peers. We all know humans with this trait, and nature is full of them. Freeloaders hide around every turn and corner of the evolutionary path. Eventually, again through some chance event, a selfish replicator would arise.

When Individuality was Invented, Sex had to be Reinvented

In his powerful essay "Tragedy of the Commons," Garret Hardin pointed out that a cheater, by taking more than his fair share of a group resource, benefits more than the amount his cheating costs any other member of the group. This is because the benefits are privatized, obtained by the cheater alone, whereas the costs commonized, shared by the group. Because of this incentive for individuals to cheat, societies look for means to deter cheaters. We put cheaters in jail and otherwise protect society from their behavior. Since sharing of proteins among the first replicating molecules would have been a kind of primitive social system, but without a means of policing the actions of cheaters, the tragedy of the commons probably existed right on life's first doorstep. This would have severely limited progress in the protein-making business. A way was needed to keep proteins from being used by selfish replicators. This led to the invention of a very special structure, an encapsulating barrier—the first cell. With the invention of the first cell, proteins could be contained, so that replicators could keep the fruits of their labor.

We cannot be certain exactly how the first cell was created. Through a series of chance mutations and recombination events, some replicators acquired the ability to construct little shelters around themselves, enabling them to hoard their own proteins for their exclusive use. Sooner or later this advantage would lead, again by chance mutation or recombination events (probably around 2,000 or 3,000 million years ago) to the creation of the first cell. What a wonderful creation this was: protection and nutrition for the replicator all under the shelter of its own roof. No longer were its proteins shared by its neighbors. With the advent of the cell, evolution would become a whole new ballgame. Individuality was born.

Individuality carried a certain cost, however. The replicator, now encased in a cell, was forced to keep both its bad mutations (improperly sequenced nucleotides) and damages (nonnucleotide intruders on the sequence). Any errors the replicator acquired—by exposure to UV light, for instance—would now be trapped in the cell. The days of free exchange and easy access to spare nucleotides were over. Sex had came naturally and without effort to the naked molecules. Now imprisoned in a cell of their own invention, the replicators had to figure out an alternative way to make repairs and new combinations. A new kind of sex between cells had to be invented.

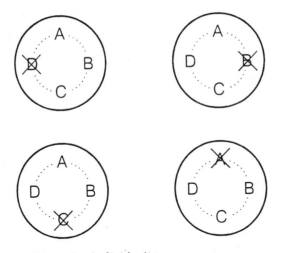

Figure 5. Individuality traps errors

What properties might these early cells have had? For economic reasons, we might imagine the early cells as carrying as few copies of each gene as possible. A single copy of each gene, as in a haploid cell, would be most efficient in terms of replication. Alternatively, for reasons of damage repair, we might imagine early cells having several copies, say two, of each gene, thereby ensuring that the cell could keep a functional set of undamaged genes. Two copies of each gene, as in diploid cells, are better than one copy; however, carrying and replicating two copies of each gene is costly. For such simple cells, the costs of maintaining and copying genes is significant. Haploid cells would, therefore, enjoy more efficient replication than diploid cells, while risking death due to gene damage. On the other hand, diploid cells, while more stable in the face of damage, would replicate more slowly and less efficiently than haploid cells. Perhaps sex, by alternating between the two types as needed, could combine the advantages of both haploid and diploid cell types while minimizing their costs.

Origin of Sex

Sex between cells typically involves fusion of the cells followed by splitting. In Figure 6, we consider sex between two gene-damaged

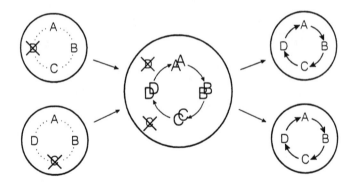

Figure 6. Sex between damaged haploid cells

haploid cells to form a temporary diploid cell, that, after repairing the damage, splits into two healthy haploid cells.

This view of the origin of sex requires that gene-damaged cells be able to carry out certain functions, such as fusion and splitting, even though their genes are damaged. Is it reasonable to assume this? We know that modern organisms and cells may carry damaged genes while otherwise appearing to be healthy. The reason that gene-damaged cells may otherwise function is that not all genes are needed in all cells all the time. A cell uses just a small fraction of its genes at any given time. Even for essential genes, their products may be stored in the cell so that the gene needn't be active all the time. Damage that occurs after a gene has been expressed will not affect the functioning of the cell unless the cell needs to copy the gene or express it again. But when it comes time to reproduce, all genes must be copied. Unrepaired damage typically blocks DNA replication so that damaged genes cannot replicate. A single occurrence of damage at just one of the many millions of nucleotide positions can keep the whole cell from reproducing. Either offspring are not produced, or if they are they are defective in some way.

What is the best strategy: diploidy, haploidy, or sex? Diploidy has the advantage of effective repair but at the cost of keeping and replicating two sets of genes (like embarking on a trip with a second car in tow). Haploidy is quick to replicate but sensitive to damage. Might sex combine the advantages of both?

To answer this question, Andy Long and I have studied mathematical models of competition between haploid, diploid and sexual cells in damaging environments. We found the critical parameters to be those describing the fusion of sexual cells, the splitting of fused cells, and the level of gene damage in the population. The outcome of competition can be represented in a three-dimensional space where the three axes represent a range of values of these three parameters, as shown in Figure 7. We found that sex may win and outcompete either type of asexual form (haploid or diploid), if sex is instigated by damaged cells.

Mathematical models cannot prove whether a hypothesis is correct. They can, however, rule out illogical and improbable ideas, as well as give insight into what factors are likely to be important in nature. The mathematical results just discussed show that our hypothesis makes sense. Sex between cells may have been invented to regain the benefits of damage-repair enjoyed by those carefree replicators. In Chapter 6, we reconsider this hypothesis by studying sex between some of the simplest living creatures on earth today—bacteria and viruses. There are other possibilities, however, as to why sex began.

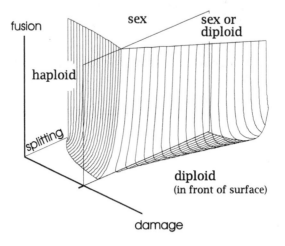

Figure 7. Origin of sex between cells

Manipulated into Having Sex

One such possibility is that sex is a vehicle for viruses and similar organisms. When two cells (or organisms) have sex, they are intimately involved to say the least. This has not gone unnoticed by selfish replicators—viruses, transposons, plasmids and the like—who have evolved ways of using this opportunity to transfer themselves from one cell (or organism) to another. Indeed, the simple bacterium *E. coli,* found in the human intestinal tract, is manipulated into having sex by a plasmid that lives inside the bacterium. Plasmids are circles of DNA that freely exist in the bacterial cell; they may hang out in the cell, they may take over the cell and divert its machinery to making plasmid babies, or they may insert themselves into the bacterial chromosome where they are copied and passed along when the bacterium replicates its DNA (the plasmid's DNA included) and divides. The fertility plasmid, or F-plasmid for short, has a number of genes that can be expressed in its host cell; there are genes for making more plasmids and genes for making the bacterium conjugate—that is, have sex—with another cell. The simple plasmid orchestrates a complex series of interactions by which its host cell makes a pilis—a tubular, penislike sort of device—that connects it to another bacterial cell that doesn't have the plasmid. The F-plasmid then copies itself and moves from one cell to the other. In the process, it may by chance drag a few bacterial genes along for the ride. This is how two *E. coli* cells have sex, cajoled and manipulated into doing so by a plasmid trying to move from one cell to another.

This is another way in which sex may have begun, not for the well being of the host cell's genes, but for the transfer of selfish genetic elements that live in the host. Though there are many aspects of sex in bacteria that this view cannot explain, it remains a possibility that it occurred during the early stages of life on earth. One could imagine how the host cell could start using the genes transferred by the plasmid for its own benefit (repair, or new gene combinations). But there is no evidence that the *E. coli* bacterium actively does this.

Genes for Breakfast

A second possibility for the origin of sex involves food. Genes are rich in sugar, phosphorus, nitrogen and other goodies that may nourish a starving cell. Some biologists believe that sexual cells were first looking for nourishment, not templates for repair or new gene combinations, when they took in foreign DNA. This idea is attractive because large amounts of free DNA exist in the mucus-rich environments in which sexual bacteria are found. In addition, sexual bacteria seem to regulate their sexuality in response to nutrients—they tend to become sexual under starving conditions. At best this idea can only be part of the answer. Although it may explain why cells would bring in foreign DNA, as occurs during outcrossing, it cannot explain why cells would go to the trouble of recombining that DNA into their own chromosome. And it is this that we wish to understand. Sex is recombination plus outcrossing. Nevertheless, the nutrition idea remains a possibility for why cells brought in foreign DNA in the first place.

Further Evolution of the Cell

With the invention of the first cell, evolution had crossed a threshold of unknowable potential. The first replicator began its journey some 4,000 million years ago, and it took about 1,000 million years for evolution to produce the first bacteria-like cell. It would be another 2,000 million years before the first eukaryotic cells (cells that make up all higher forms of life, in which DNA is packaged in a nucleus). Most animals and plants evolved rather recently, during the past 1,000 million years. More than 3,000 million years, over three-fourths of life's history, was spent producing the cell, the basic unit from which all higher forms of life are built. Once invented, these cells could combine with one another into ever more elaborate and intricate patterns, creating organisms large enough to be seen by the naked eye.

So it is at this point in the history of life that we must drop our magnifying glass, step back from the primordial pool, and watch as the seas fill with living forms, some strange and others familiar from our modern vantage point. Out of this plenitude, a funny-looking fish, with deformed fins inappropriate for swimming but able to support it on land, would emerge. It would crawl out of the sea that nurtured all life for some 3,000 million years and, making use of its novelty, strike out alone. This animal would spawn an explosion of new forms of life, some more and some less dependent on the nurturing sea. Eventually, land-dwelling forms would evolve ways of taking the sea with them by first packaging water in cells and later by inventing circulatory systems that bathed each cell in a sea-waterlike fluid (blood). These inventions would allow life to colonize new habitats on land and in the air, far away from the sea. Finally, having produced an erect, bipedal ape, too large to avoid its predators yet too small to fend for itself, trapped and exposed in an open savanna, evolution would experiment and, for the first and perhaps last time, produce a thinking, conscious being. In time, this being would come to wonder about itself, its origins and nature.

◆ 3

Mrs. Anderson's Baby

The garden flew round with the angel,
The angel flew round with the clouds,
And the clouds flew round and the clouds flew round
And the clouds flew round with the clouds.

Is there any secret in skulls,
The cattle skulls in the woods?
Do the drummers in black hoods
Rumble anything out of their drums?

Mrs. Anderson's Swedish baby
Might well have been German or Spanish,
Yet that things go round and again go round
Has rather a classical sound.

—*Wallace Stevens*
"The Pleasures of Merely Circulating"

Life on earth can never be repeated. If life were to begin again on this planet, even given identical circumstances, chance events would lead it down a different path. The evolutionary process opportunistically evaluates random events (such as mutations and new recombinations) and channels them into useful structures. Some broad patterns might be the same—flying animals would still need winglike structures, swimming animals would need fins and some kind of sex would likely exist—but individual species would certainly be different the second time around. Human beings would probably not exist.

33

The World Might Have Been Otherwise

Our world, our species, each one of us . . . might never have been. This may be the most difficult lesson the theory of evolution has to teach us. At first it may seem threatening to consider the idea that we owe our very existence not to the inspiration of a good and omnipotent being but rather to natural process operating on chance events—evolution. It makes one feel, well, a bit lonely. Comfort and inspiration can be found in the realization that all living things are one—that, in a real informational sense, the living world is one big family. Other cultures (those of Native Americans come to mind), have embodied this attitude about the living world in their myths and teachings. Unfortunately, the Bible got humankind's place in nature completely wrong. Instead of being a quantum leap above animals as proclaimed in Genesis, our species is found on a tiny twig, on a rather short branch, off to the side of the great tree of life.

Perhaps this is a major reason why many people do not believe in evolution. It certainly can't be because the theory of evolution is technically difficult or complex like, say, the theory of relativity. The basic idea of natural selection is simple and was introduced in the previous chapter during our consideration of molecular replicators. Apparently, it is easier for some to believe in mystical forces and unseen events than to accept Darwin's simple but compelling logic. Others understand Darwin's logic but are unwilling to embrace its profound implications. It has been some 130 years since the *The Origin of Species* appeared, yet the Darwinian revolution has barely begun.

Evolution is a Fact and Evolution is a Theory

Those who find evolution threatening often find consolation in the notion that evolution is, after all, only a theory, not a fact. (Indeed, two recent presidents of the United States, Ronald Reagan and George Bush, fall back on this interpretation when asked their views on evolution.) The implication is that evolution might not be true. To think this way is to be confused about the role of theories and facts in science.

Evolution is both a theory and a fact. The fact of evolution—that species have evolved from one another over billions of years—is indisputable. The fossil record, comparisons between living species, and, more recently, analyses of the real book of instructions—the nucleotide sequences of the DNA molecule itself—indicate clearly that all living things, ourselves included, are related and have evolved from earlier forms of life. These are some of the facts of evolution.

The exact way in which evolutionary change has occurred is, in some cases, contentious and open to several interpretations. For instance, there is no debate in the scientific community about whether humans are closely related to apes in the evolutionary tree (this has been known for more than a century), however, there is disagreement over whether humans are more closely related to chimpanzees or to gorillas. The resolution of this particular issue doesn't bear on whether evolution has occurred, for we know it has, or whether humans, chimps and gorillas all share a common ancestor, for we know they do.

To what then does the *theory* of evolution refer? To the set of hypotheses by which the facts of evolution are explained. The Darwinian theory of evolution shows how the various forces of natural selection, genetics, ecology and history fit together to explain the observed patterns. Foremost among the forces used to explain evolution is natural selection. We were introduced to this process in the last chapter when we considered the early replicators. Evolution by natural selection is the unavoidable consequence of three properties of life: variation, heritability and what Darwin termed the "struggle to survive."

Natural Selection

Darwin observed that all living things have the potential to increase in numbers at an exponential rate. In the case of the elephant, one of the slowest-breeding animals, he calculated that if each breeding pair had but six surviving offspring between the ages of 30 (when elephants first begin reproducing) and 90 (when they die) there would be more than 15 million elephants descended from the first pair in just 500 years. Faster breeding creatures would fill the earth in only several hundred years. Obviously, only a few of these offspring actually survive. Consequently, there must often be a struggle to survive and reproduce by which organisms that survive to reproduce overcome any threats to their existence—bad weather, food shortages and predators, to name a few.

No two organisms are identical, and thus they are usually not equally able to survive and reproduce. Due to the traits they possess (and, ultimately, to the genes that determine these traits) some kinds of organisms are more able to survive and reproduce than other kinds. Those ancestors of the present day giraffe that had longer necks than their peers, for example, were better able to withstand an environment in which food was to be found high in trees. Such variations in "fitness" provide the raw material for evolution by natural selection so long as they are heritable, that is, capable of being passed on from parent to offspring.

Darwin knew very little about how genes work. But he did appreciate that offspring tend to resemble their parents and, furthermore, that this resemblance was necessary for evolution by natural selection. Today we know that heritability results from the faithful and accurate replication of the nucleotide sequence information contained in DNA. Offspring resemble their parents because they possess copies of their parents' genes.

Putting together the three concepts of variation, struggle to survive and reproduce, and heritability, we see that heritable traits that aid in the struggle to survive and reproduce should increase in frequency in the population. This is because the carriers of such traits have more offspring and these offspring carry the genes for the traits. It is this steady increase in the frequency of beneficial traits and the complementary decrease in the frequency of disadvantageous ones that Darwin called natural selection.

Individual organisms do not evolve. Only populations—groups of organisms that belong to the same species—evolve. Organisms are born and they die. Between birth and death, they may have offspring, and some may have more than others depending on the traits they possess. These different rates of reproduction cause the *population* to change its genetic composition—that is, to evolve. Over time, for instance, if the long-necked giraffes reproduce in greater numbers than the short-necked giraffes, the length of giraffes' necks will increase even though each individual giraffe had a fixed neck length during its adult life.

Darwin's genius was in realizing that all the apparent perfection of design in the living world resulted from this simple mechanistic process. Before Darwin, people of the Judeo-Christian or Moslem traditions had assumed that living things must have been created by a being such as God. Darwin showed that the creator of all living things was, in fact, not a divine being who stood apart from life, but a *process* intrinsic to life—natural selection. Ultimately, this process provides the rationale for our existence, and it is in this process that we must seek an explanation of our basic nature, the origin of our needs and desires, and the purpose of sex.

Life Could Not Exist Without Evolution

Life and evolution are inextricably linked; one could not exist without the other. Those who claim to respect life but who do not believe in evolution will never understand the nature of life. Living things—butterflies, robins, roses, ourselves—are temporary repositories of genetic information, information that is in the process of being shaped and molded by evolution. If the evolutionary process were somehow frozen in time, organisms would soon decay and disappear. Woody Allen said in the movie *Annie Hall* that human relationships are like sharks: if they don't keep moving forward, they die. What is true of vitality in human relationships is even more true of the continued vitality of life itself. Life must continually change—it must evolve—to exist at all.

Evolution: The Science of "Why" Questions

In attempting to understand why sex exists, we will ask how sex works. In the nonliving world of physics and chemistry, it is usually not useful to distinguish between *how* and *why* questions. When we ask why a rock falls down into the sea, we really mean to ask how did it occur. We are satisfied with an explanation in terms of the proximate forces of gravity, friction and the rainstorm that loosened the rock's footing. We are not tempted to wonder why the rock exists, why physical forces exist or why the storm happened to rain at that place at that particular time. Of course, we might wonder about these matters, but they will not help us to answer the first question.

But organisms are purposeful, their purpose being to pass on their genes. Once we understand how a particular feature or behavior of an organism works, we are naturally led to wonder why the feature or behavior exists. We often ask, in what way does the feature or behavior contribute to fitness? In biology, it is useful to separate the two kinds of explanations, the how and the why, while keeping in mind their interdependency. Were we to ask, for example, as we will shortly, why males and females behave so differently, we might take two distinct approaches in our answer. We might point to the different levels of various hormones in males and females, or to differences in the structure of their brains as possible explanations. Such explanations would really be answers to the "how" questions. How is behavior controlled and what are the proximate factors involved.

In this book, we are primarily interested in "why" questions. In the example just discussed, we will ask why the male and female brain and endocrine system are organized so differently and why the two sexes look and behave so differently. Evolutionary biology has a method for dealing with such why questions. This method involves understanding the heritabilities of the traits considered and whether they are correlated with other traits affected by natural selection. It requires that we understand male and female ecology, and the different strategies males and females pursue to successfully reproduce and pass on their genes. Once natural selection has been brought to bear on the why question, our pursuit of the why can rest, for we know why natural selection occurs. It is the unavoidable result of differences in fitness and heritability. The fruitful pursuit of the why stops in evolutionary biology. There is no place else to go.

Understanding why a trait exists often requires understanding how organisms work. The "how" question suggests answers to the why question. For example, if we were to study swimming behavior in fishes and come to wonder why different fish have different body shapes, it is useful to know something about the hydrodynamics of movement in water—what physical forces are at work as a fish of a given shape moves through the water. Likewise, to answer the question of why sex exists, we need to know how sex works at the level of DNA. This will lead to what may appear to be an overly technical diversion into recombination enzymes and chromosomes in Chapter 6 and selection mechanisms in Chapter 7. Please bear with me if you find these parts difficult.

Not all human, let alone biological, phenomena need have "why" answers rooted in evolutionary biology. Indeed, when the topic is human behavior, many scholars believe just the opposite. When asked why you prefer Mozart and I Bruce Springsteen, for example, we don't seek an explanation based in natural selection. But if we were to wonder why humans appreciate music, or even more generally, why they have any esthetic sense at all, then we might consider possible selective pressures that might have operated among our ancestors to promote this capacity. We might also investigate whether the musical sense is a by-product of natural selection acting on other mental capacities.

The problem is that it is not always clear when the why questions should stop. It could be that you and I prefer different music because our ancestors lived in different environments and experienced different selection pressures, one set leading to a preference for classical music and the other for rock and roll. It's not likely that this is the case, though, because one's taste in music usually does not affect survival and reproduction (although it is possible to imagine environments in which it could). Furthermore, there is likely little, if any, genetic basis to music preferences. In other cases, however, it is more difficult to know when the why questions should stop, and this can lead to endless speculation concerning invented "just so" stories, like Dr. Pangloss's observation in Voltaire's *Candide* that humans have noses so as to wear glasses.

Thinking Evolutionarily About Sex

In the case of sex, there are a variety of reasons that give us license to pursue the why approach back to evolutionary origins. First, sexual behavior has direct and immediate consequences for individual fitness. It affects such key components of fitness as our chances of mating, who we mate with and, since sex often involves taking risks, our chances of survival. Although humans have lost the capacity for virgin birth, asexual reproduction would have been a possibility in our distant ancestors. Among these ancestors, the number of offspring produced and their survival would have depended on whether they were produced sexually or not. We therefore expect natural selection to have paid special attention to sex.

Second, there is great variation in sexual behavior. There are the obvious and not-so-obvious differences between male and female sexuality, and within each gender there is also much variation in sexual behavior and attitude. How much of this variability has been shaped by natural selection and how much is incidental is an open question. (From the point of view of natural selection, any difference that is not inherited is inconsequential noise.) But we can assume that the basic differences in sexual behavior between males and females are based to some extent on the expression of different genes. Although the genomes of males and females are much the same, differing only in that males have an X and a Y chromosome and females have two X chromosomes (there are only a few genes on the Y chromosome), there are significant differences between males and females in which genes are expressed. Many genes, for example those coding for the reproductive organs or mammary glands, although present in both sexes, are expressed in only one sex and are silent in the opposite sex. The expression of these sex-specific genes is triggered by the presence or absence of the Y chromosome.

When it comes to sex and sexual behavior, both conditions necessary for natural selection—variation in fitness and heritability—are met. So, it should be fruitful to consider the evolution of the sexes and sexual behavior. What does it mean to be a male or a female? This question obviously has many facets, including the psychological, physiological, morphological and hormonal ones. We are primarily interested in the most basic of biological differences involved and why evolution produced them. Why did evolution invent males and females in the first place? As the biblical story of creation tells, the male came first and the female was created so that he would not be alone. What we know about asexual reproduction and virgin birth in other animals suggests the opposite: that females, or at least the female function, came first and can exist without the male function. We will find that males have no biological purpose without females, but females may exist without males; which is, after all, what asexual reproduction is all about. But what is the male and female *function*, what is the purpose of the male and female sex?

Definition of Female Sex

We are so caught up in our own biology and psychology that it is difficult to appreciate the original purpose that males and females had. To help us stand back and gain perspective on the confusion, let us return to the egg and the sperm cell. Even at this level there are marked differences between males and females in our species. The egg is large in size (over 1,000 times the size of the sperm), immobile, and rich in nutrients and other resources needed once it is fertilized and begins developing into a new organism. The sperm, on the other hand, is small, designed for swimming using its tail, and devoid of resources. Even the energy it needs for swimming is provided by the parent cells that produced it (enough for a few hours of frenzied pursuit of the egg).

In simple sexual creatures, like yeasts or the green algae *Chlamydomonas*, two kinds of gametes are produced that must fuse with one another, but they are not different in size, shape or function (except that fusion involves one of each type). In the absence of any gamete dimorphism, the words "male" and "female" are not used; instead the two gamete types are referred to with more neutral terms like "+" and "−" or "a" and "α." We use only the words "male" and "female" when the gametes are of different sizes; when this occurs, the large gamete is referred to as "female" and the small gamete as "male." Biologists use the term "isogamous" to refer to sex between similar gametes (as in *Chlamydomonas* or yeast); we use the term "anisogamous" to refer to sex between dissimilar gametes (as between egg and sperm cells). Whether sex is isogamous or anisogamous has profound consequences on the cost of sex.

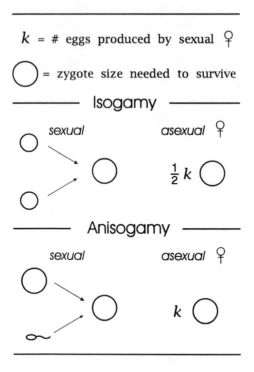

k = # eggs produced by sexual ♀

◯ = zygote size needed to survive

——— Isogamy ———

sexual   asexual ♀

$\frac{1}{2}k$

——— Anisogamy ———

sexual   asexual ♀

k

Figure 1. Cost of sex and gamete size

No Cost of Sex for Equal-Sized Gametes

Let us reconsider the cost of sex—more specifically, the cost of males—in a little more detail. Consider Figure 1, a diagram of the two kinds of sex: isogamy (same-sized gametes) and anisogamy (different-sized gametes). We assume that the survival of the zygote depends upon its size, zygote size being a measure of the nutrients received from the parents (of course, there is just one parent in the case of asexual reproduction). The size needed for the zygote to survive is shown in the figure. In anisogamous sex, the mother's egg contributes all the nutrients and volume of the zygote, since the father's sperm contributes next to nothing except genes. The number of eggs produced by a mother will be the same whether she is sexual or asexual, since both kinds of mothers must produce eggs of the required size (or else their offspring will not survive). Consequently, an asexual mother will produce the same number of zygotes as will a pair of sexual parents, mother plus father (*k* in the figure). However, all of the asexual mother's zygotes will become egg-producing females like herself, whereas only half of the zygotes produced by sexual parents will become egg-producing females (the other half become fathers). After a single generation, the number of zygotes produced by asexual females will be twice as great as those produced by the sexuals, all other factors being equal. This is the twofold cost of males mentioned in the first chapter.

If sex involves the fusion of isogamous gametes, there is no cost of sex. The only cost of isogamous sex is the time and energy it takes. In isogamous sex, each of the parents contributes equally to the zygote, approximately one-half the zygote's volume coming from each gamete. An asexual parent must produce the whole gamete itself, so its gametes must be twice the size of those of the sexual parents. Consequently, asexual parents produce only half as many zygotes as do sexual parents. After one generation the number of zygotes produced by the asexual parent is the same as that produced by sexual parents, all other factors being equal. Thus, there is no cost of sex other than the time and energy needed for fusion and splitting.

Origin of Sex Differences

Anisogamous sex is by far the most usual kind of sex among complex organisms. Why? Understanding the origin of anisogamy will help us to understand the biological nature of males and females. Sexual reproduction requires that the parents find each other so they can mate and the offspring be cared for. These two responsibilities, mate finding and nurture, are basic to the success of sexual reproduction. Under a wide range of circumstances it is difficult for cells to do both, that is, it is hard to be both rich in nutrients and mobile. Biologists believe that this simple constraint underlies the basic differences between eggs and sperms. Eggs are specialized in the nurture function and sperms are specialized in mobility and the mate-finding function. Together two specialized gametes (anisogamy) are better able to produce a healthy zygote than are two identical gametes (isogamy).

Many of the behavioral and morphological differences between males and females stem from the often conflicting needs of nurturance and ability to get mates. The female sex typically specializes in nurturing the zygote while the male sex typically specializes in finding females. As a result, the egg is far more costly to produce than the sperm. Having invested so much in producing an egg, a female is less inclined to abandon her offspring—if she does, she will have to go through the costly process once again. In contrast, a male invests only a rather cheap sperm in the zygote, and so is more inclined to abandon it and try to fertilize another egg. Since the female sex is defined as the one with the large gamete to begin with, she is usually the parent that cares for the offspring. There is much more to gender differences than this; however, the needs of nurturance and mate finding are still the dominant themes in understanding these other differences.

Consider two other gender differences as examples: female choice and male displays. It is usually the female sex that chooses to mate males being willing to mate with almost anyone and anything resembling a female of their species. It is usually the male sex that is adorned with beautiful coloration and displays. These differences result from a special kind of natural selection that Darwin termed "sexual selection." Sexual selection is the selection that occurs with regard to features of the organism involved in mating, features like differences in body size and weight, bird song or plumage coloration, or even bizarre features like the male peacock's tail or the large rack of antlers that male deer have. These features are explained by evolutionary theory in terms of their advantage in mating, even though, like the huge rack of antlers of the Irish elk, they may handicap the male's survival. Such bizarre features are testimony to the significance and power of the mating game and more evidence that sex is costly.

Sexual Handicaps

How can evolutionary theory explain such costly sexual handicaps? There are two basic ideas. The first is referred to as the "runaway process" and was proposed by the Englishman Ronald Fisher, a name we will encounter many times in our discussions of the evolution of sex. The basic idea is that females prefer to mate with males with a certain trait, which we will refer to as the male display. Perhaps because of the way the visual system is designed, females are more able to see a male with a colored tail, so that a colored male gets more matings. The more extreme a male is, for example, the more brightly colored he is, the more matings he gets—passing on his genes for bright color, or whatever the male trait is, to his offspring in the process. If there is any variability for choosiness among the females (perhaps some females are more able to see bright colors than others) then these genes for preferring brightly colored males will become associated with the more extreme color causing genes. This genetic association (termed "linkage disequilibrium" and discussed in the notes for Chapter 5) generates positive feedback in the selection process and acts to produce ever choosier females and ever more extreme males. At some point, the costs of females becoming choosier and/or males developing brighter or larger displays stops the runaway process and the species is left with extremely colored males preferred by choosy females. This logic can apply to other male traits, such as body size, degree of ornamentation or the size of a buck's antlers. In spite of initial widespread acceptance of the runaway process, recent mathematical and computer models have shown that the positive feedback generated by the genetic associations (the previously mentioned associations between the genes causing a female to be choosy and those causing a male to be more extreme) is too weak a force to drive the process to the extreme values observed in nature.

There is nothing adaptive about the runaway process from the point of view of the species' ability to persist in its environment. The brightly colored sons of choosy females get more matings, true, even though they may have a hard time surviving because of the cost of their display. However, the daughters of choosy females are not any better off for having a brightly colored father. The second idea about sexual handicaps focuses on the quality of the daughters and argues that daughters of choosy females *are* better off because their brightly colored father has better genes than the drab males in the population. In other words, females prefer extreme males *because* the male's displays are indicators of genetic quality. This idea is termed the "handicap principle" and was proposed in the late 1970s by the Israeli ornithologist Amot Zahavi.

What do we mean by genetic quality? We will define genetic quality as the viability an organism would have, male or female, if we could ignore the costly effects of the display (in the case of males) or the costs of being choosy (in the case of females). Genetic quality refers to fitness in the absence of the traits we are interested in understanding (male display and female choice). There will always be differences in genetic quality in the population, due perhaps to deleterious mutations which are unavoidable. Some males will have more, some less, numbers of deleterious mutations because of the recurring and random nature of mutation. Wouldn't it be nice if a female could pick out from this variability in quality the best male, the one with the fewest mutations. The central assumption in the handicap theory is that a male's display advertises his genetic quality; males with bigger displays have better genes. But what keeps the display honest?

Honesty in Advertisement

The display must be an honest advertisement, otherwise females would lose interest with time. What is it that keeps poor quality males from having a large display? For example, why can't a low-quality stag have a huge rack of antlers with many points? Zahavi's answer is that males of different qualities are more or less sensitive to the cost of the display. The display is a handicap and only high-quality males can pay its cost. Consider males of high and low quality (ones with many or few deleterious mutations) and a particular value of the display, say a rack of antlers with a certain number of points. High-quality males bearing this display will have higher survival rates than low-quality males bearing the same display.

Now consider a range of values for the display from small to large. Survival for both low- and high-quality males will decline

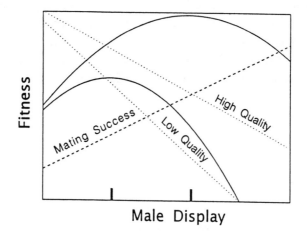

Figure 2. Handicap principle

with size of the display, but the survival of low-quality males will drop off more steeply (shown by the dotted lines in Figure 2). Mating success for both kinds of males goes up with the size of the display (shown by dashed line). The overall fitness of the two kinds of males is the product of their survival and mating success, shown by the humped curves in the figure. The overall fitness of the two kinds of males is maximal at an intermediate value of the display (the peak of each curve). However, the optimal value of the display for the high-quality male is greater than that for the low-quality male. It doesn't pay the low-quality male to lie and produce a display any larger than his optimum. The cost of the display keeps male advertisement honest. More explicit genetic models of the evolution of the display have confirmed this heuristic reasoning.

A Bumpy Ride

These brief excursions may have left you feeling a bit disappointed in the whole enterprise and thinking that evolutionary biology does nothing more than justify certain traditional stereotypes of male and female roles. Keep in mind that we are trying to understand our biological nature and evolutionary history, not how males and females *should* behave. The way the world happens to be (say as a result of evolution) may not be the way we want it to be; in philosophical jargon "is" does not imply "ought." If nature has endowed us with predispositions to behave in ways we don't like, all the more reason to work hard to change them. Evolution has also endowed us with enough flexibility to try to do that.

While we are on the subject of what comes naturally, evolution is often said to be "selfish" in the sense that organisms who benefit themselves at the expense of others often enjoy an advantage in the struggle to survive. What is commonly not appreciated is that the advantage to selfishness occurs only in a certain kind of population, namely one that is extremely large and well mixed, so that organisms do not encounter relatives or do not meet with non-relatives more than once. However, these conditions are not often met; populations are small or they are structured into groups of relatives or clusters of non-relatives who interact repeatedly. In these situations selfishness doesn't pay; cooperation, even altruism, between organisms can be favored over selfishness by natural selection. So, selfishness between organisms is not always "natural." The "is" as a result of evolution may often be cooperation. My point is simply that there is nothing "unnatural" about helping others.

The history of life has been a bumpy ride, not a trip well planned in advance. It has depended on chance variations and fortuitous events. It is not repeatable and would certainly happen differently a second time. The theory of evolution is the best explanation we have for how it happened, but it is not based on a system of moral beliefs. To use its capricious inventions as substitutes for moral values would be a mistake—a great mistake indeed. Evolution is simply too capricious and undependable for that. As with Mrs. Anderson's baby in Wallace Stevens's poem "The Pleasures of Merely Circulating," the world could have been an entirely different place.

We have discovered that differences in sexual behavior are rooted in the origin of the male and female sexes; the origin of the male and female sexes stems from a simple constraint inherent in sexual reproduction, that large gametes tend to be immobile. The problem of the origin of sex has already taken us back to the origin of life itself (Chapter 2) and will take us into the realm of death and immortality in the next chapter. In later chapters, sex will lead us to question the existence of organisms as meaningful objects in evolution (Chapter 9) and to wonder about the landscape and organization of variability in the living world (Chapter 10).

Sex and Death

The object of love, Socrates, is not, as you think,
beauty. . . . Its object is to procreate and bring forth in
beauty . . . because procreation is the nearest thing to per-
petuity and immortality that a mortal being can attain.
—Diotima in Plato's
Symposium

The flower of the century plant is so huge, over ten times the size of the basal plant, that it looks more like a tree than a flower as it towers above the sparse Sonoran desert landscape. The century plant waits quietly, ten, twenty, sometimes fifty years, as a rather dull rosette, a clump of spiny shiny leaves, before it shoots its spectacular flowering stalk towards the sky. Soon afterwards it shrivels and dies. Biologists call it "big bang" reproduction. The little Australian marsupial mouse *Antechinus* dies immediately after a single episode of frenzied mating. Or consider the poor praying mantis father; no sooner has he finished with the pleasures of mating, than he is eaten whole by his mate. Is it our Victorian outlook, or is sex really bad?

Sex is for Genes and Genes are Forever

These examples may sound extreme, but they illustrate the kind of trade-offs between reproduction and survival that inextricably link

50

sex with death. In the first chapter, we introduced the costs of sex, and the costs of males, that stem from parents producing offspring sexually. These costs involve the genetic costs of sharing genes with a mate, of producing male offspring that do not produce eggs, as well as the costs of mating. The costs of mating are complex and varied. One example is the cost of finding mates when the species is rare, and this will be the subject of Chapters 9 and 10. The costs we are considering now are more direct: the time, energy and effort devoted to keeping and finding mates, as well as to mating and to reproduction itself.

It's hard to do more than one thing at a time and do it well. One of my first shop tools as a teenager was a jack-of-all-trades device that, by unbolting and turning some levers, could become a drill press, a lathe, a table saw, or a radial arm saw. It didn't perform any of these functions particularly well, but it was all that I could afford (plus, it was small enough to fit in my parent's basement). Each function seemed to work against the other. The long reach needed for the drill press or lathe made for an awkward table saw.

And so it is with organisms. The long and brightly colored tail needed to attract a female peacock makes for awkward goings in the weeds and brush. The well-nourished fetus puts mother at all kinds of risks; late in pregnancy it is even difficult for her to walk. Imagine our ancient grandmothers trying to climb a tree to avoid a tiger on the open savanna with a 9-month fetus inside her. For that matter, the modern mother-to-be is not much better off when crossing busy West 53rd Street in Manhattan.

Then as now, babies are a burden. Mutant female fruit flies without ovaries live longer. Even sperm, often ridiculed because it is cheap and poorly constructed (and when compared with the egg, it is), lowers survival when it is produced by males in the simple parasitic worm *C. elegans*. Reproduction, sexual or asexual, involves risks both large and small. A mother whiptail lizard, although able to reproduce without a male, must still allocate a great deal of time and energy to nourishing and guarding her eggs—time and energy that could be spent on finding food for herself. But at least she doesn't have to bother with a mate. Asexual reproduction is hard enough and sex only further complicates matters, for it requires coordinating the lives of two entirely different and usually unrelated individuals. Finding, attracting and keeping a mate adds greatly to the already costly task of reproduction.

For individual organisms, sex is not just a bother, it is downright dangerous. If not the harbinger of death—as it is for the century plant, the praying mantis and *Antechinus*—sex hastens the journey to the grave for all of us. If the direct costs of energy and time don't get you, the parasites and infectious elements that, like the AIDS virus, have seized this chance to transfer themselves will. In spite of the immense pleasure and satisfaction we derive from it, sex is not good for us. Nor is it meant to be—as we know, sex is for our genes.

All organisms must die, of course. Being born, dying, and having offspring in-between: this is the biological imperative. Reproducing is, after all, what organisms are about—genes invented organisms as a means for their own perpetuation. We will find in the chapters that follow that as far as the perpetuation of genes is concerned sex is a good thing, cleansing the genome of life-stopping damage, purging deleterious mutations, perhaps even providing hopeful variations for new environments. Although organisms come and go, genes are forever.

Germ Line

In humans, and other organisms with many differentiated cells, sex culminates with the fusion of two cells—an egg and a sperm. The newly fused cell, the diploid zygote, undergoes a few mitotic divisions until a fundamental split occurs in the cell line. One line of cells, the "somatic line," continues dividing by mitosis to produce all the organs, tissues and other somatic cells that make up the organism. This somatic line of cells is mortal and eventually dies out with the death of the organism. Another line of cells, the "germ line," separates off from the somatic line early in the life of a developing individual and temporarily stops dividing after a few divisions. Once sexual maturity is reached, germ line cells continue dividing mitotically for a few more divisions but culminate in a final meiotic (sexual) cell division to produce the specialized haploid sperm or egg cells. It is during this final meiotic cell division that recombination occurs at high frequency to prepare the cell's DNA to be passed to the offspring by cleansing it of life-threatening damage. In stark contrast to the mortality of the somatic cell, the germ line is immortal.

Germ line cells belong to a distinguished line of cells that can be traced backward and forward in time, for all time. Backward in time, through parents, grandparents, great-grandparents, through all recent relatives and ancient ancestors. Back 1½ million years, through those big-brained bipedal hominoids that link our lineage, *Homo,* with the now-extinct *Australopithecus.* Back 6 million years, through tribes of arm-walking apes that connect us with modern chimpanzees and gorillas. Back 50 million years, to a tree-dwelling catlike prosimian primate—who held our human destiny in its grasping hands. Back 80 million years, to that little shrewlike animal that unites all primates. Back 200 million years, to that now-extinct mammal-like reptile *Theriodontia.* Back to the seas and our fish ancestors, who could be found swimming there some 400 million years ago, having just invented a marvelous internal skeleton. Back, far back, to creatures like a small wormlike marine animal that unites our ancestors to sponges and starfishes, or a single celled amoeba-like protist that unites all plant and animal life. Creatures so strange and distant that our family ties with them seem more like a metaphor than concrete genetic kinship.

But kinship it is. The German biologist August Weismann first used, in the latter part of the last century, the words "germ plasm" to describe this line of cells. "Germ," from the Latin *germen*, means sprig, offshoot, sprout, bud or embryo. "Plasm," from the Greek *plassein*, means a forming substance. Real, physical, informational connections through time: that is what the germ line is about. The germ line, that ever-branching, ever-reticulating web of cells, connects all past, present and future living things. Life originated just once on this planet, and since that time, some 4,000 to 5,000 million years ago, it has been handed down like a family heirloom, from parent DNA strand to daughter strand, from parent cell to daughter cell, from parent organism to offspring, through our bodies, out of our bodies, into other bodies, through the eons.

The egg and sperm: two cells whose job is to link the generations. Full as they are with the potential to create a new and unique individual, we wonder: To whom will they eventually lead? To what unknown and not-yet-designed organism will they eventually connect us with? Who could have thought, certainly not our distant shrewlike ancestor (probably one of the smartest creatures around at the time), that in just 90 million years of evolution there would be a species that included Albert Einstein, Vincent Van Gogh, and Margaret Mead! Nor can we know what inventions of form and function the millennia will bring forth. Yet we—which is to say our genes—are in touch with these distant ancestors through the germ line.

Many multicellular organisms, especially plants and many invertebrates (such as sponges and starfish) do not have specialized germ line cells that are kept separate from somatic cells. In these organisms, almost any cell has the potential to be a germ line cell. Of course, in single-celled organisms (protists, bacteria, and viruses) there is only one cell to speak of.

Single-celled protists, like a well-studied species of paramecium, can reproduce asexually or sexually according to environmental conditions. The rate of cell division of a lineage of paramecia slows down and ultimately dies out if the cells are forced (by environmental conditions) to reproduce asexually. The death of an asexual lineage of paramecia is like the death of the soma, the somatic cell lineage, in a multicellular organism. If the paramecia have sex, even if it is only periodically, the cell lineage is rejuvenated and continues dividing. These single-celled protists can go on dividing, apparently forever, only if they have sex periodically.

The contrast between a newborn's youthfulness and its parents maturity is as remarkable as it is familiar. It led early biologists to believe that sex somehow rejuvenates life. We have mentioned how Weismann so completely destroyed this idea, or so he thought, by countering, "Twice nothing cannot make one." So, he had looked elsewhere for an advantage to sex and found it in genetic variability.

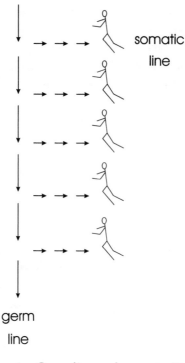

somatic
line

germ
line

Figure 1. Germ line and somatic line

But why are babies young? Something must occur in the production of the egg and sperm, something to rejuvenate the cells of an aging adult.

Aging has a Purpose

Aging is the inherent and progressive decline in the ability of organisms to survive and reproduce—in short, a decline in the ability to function. Aging does not refer to the obvious and necessary fact that with time organisms die, just as objects break and machines wear out. Most of the drinking glasses on my kitchen shelves were bought rather recently; there are very few left from the set I first bought when I went to college. Similarly, in growing populations of organisms, there will be more young ones than old. This is wear and tear, not aging. Aging is an irreversible process inherent in the genes.

Organisms die even if there are no outside disturbances of any kind. Why?

There is an advantage to having offspring as early in life as possible. The sooner offspring are produced the better for propagating genes. It's like money in the bank: the sooner you deposit it, the faster it starts earning interest. "Interest" in this case means gene copies in grandchildren, great-grandchildren, great-great . . . —you get the picture. There are more young individuals than old individuals in a growing population (for the same reason that most of the glasses on your kitchen shelves were bought recently), and genes whose effects occur early in life are under more scrutiny than are genes that are expressed late in life. The more technical way of saying this is that the force of natural selection declines with the age at which a gene has its effect.

However, a gene often has several effects (this is termed pleiotropy) and sometimes one effect is beneficial while another is deleterious. For example, the sickle-cell gene causes human red blood cells to take on a sickle-like shape, with two very different consequences. The sickled blood cells are more resistant to the parasite that carries malaria, at the expense of a decreased capacity to carry oxygen. Of special interest to the question of aging are genes that have opposite effects at different ages on their carrier's fitness. A gene that allocates more resources to reproduction early in life may leave fewer resources for repair, resulting in body tissues that are more likely to fall apart later in life, that is senesce. These early-acting genes are precisely the kinds of genes that are under more intense natural selection.

Genetic Error Causes Aging

Aging of the soma, the body tissues, is a direct result of these two aspects of natural selection and genetics: the force of natural selection declines with age and genes can have pleiotropic effects, benefiting the organism early in life but harming the organism latter in life. These are the evolutionary reasons why organisms age. But what are the underlying mechanisms? Genetic errors and their repair (or lack thereof) play an important part.

Genetic errors, both damage and mutations, accumulate in an

organism's DNA throughout its life. Mutations occur primarily when cells and their DNA are replicating, while DNA damage occurs even in the absence of DNA replication. Eventually, genetic errors accumulate to such an extent that vital functions deteriorate and the organism dies.

Researchers have found that UV radiation, a common DNA-damaging agent, significantly reduces clonal life span in simple single-celled protists, such as *Paramecium tetraurelia*. Selection among members of the clone cannot keep the lineage free of genetic error as discussed below. Particular kinds of DNA damage can be detected by geneticists and have been shown to accumulate in asexually propagated lines. Accumulation of DNA damage with age has also been found to occur in the soma of multicellular organisms.

Coping with Genetic Error

Chapters 5, 6 and 7 will elaborate on four ways of coping with genetic errors: avoiding them, selection, hiding or masking them, and repair. Each strategy is most effective in certain kinds of tissues and in certain kinds of organisms.

Selection is an especially effective way of coping with both damage and mutations in populations of replicating cells, such as in tissues of multicellular organisms or in populations of single-celled organisms. For example, blood-forming cells and the epithelial cells that line the intestines are constantly dividing and being replaced. If a cell with a large dose of damage or mutation dies, a less-loaded cell may replicate to replace it. Cellular selection is like natural selection but occurs among cells within tissues in an organism (instead of among organisms in a population). Epithelial cells don't seem to age, probably as a result of cellular selection. Likewise, plants may live for a very long time—for example, the Great Basin bristlecone pine can live for over 4,000 years. The creosote bush, common in the deserts of the southwestern United States, Central and South America, lives for around 100 years, but parts of the bush may break off and continue propagating as clones in the surrounding desert soil. A group of such clones has been found in the Mojave Desert dating back 11,700 years! Similarly, a population of dividing bacteria, like the common gut-living bacterium *Escherichia coli*, doesn't appear to age; actually, aging does occur for a single bacterium and its descendants, but natural selection keeps the population of cells relatively free of genetic error, so the population of cells doesn't age.

In multicelluar organisms, some tissues differentiate to such an extent that their cells stop dividing, and, not surprisingly, these tissues show the greatest degree of aging because cellular selection can no longer occur. For example, cells belonging to the liver, brain or muscle tissues do not divide once they are fully differentiated. DNA damage accumulates, protein synthesis and transcription of DNA into RNA decline, and the tissue gets old.

DNA repair is also an effective way of coping with genetic damage and postponing aging. Numerous studies have shown that organisms with long life spans typically have greater capacity to repair DNA damage than do organisms with short life spans. In addition, humans with diseases that have features of aging—diseases like Down's syndrome, Cockayne's syndrome, Werner's syndrome, Hutchinson-Guilford progeria, ataxia telangiectasis, Huntington's disease, xeroderma pigmentosum, diabetes mellitus, Friedreich ataxia, Alzheimer's disease and Parkinson's disease—typically have greater levels of DNA damage and/or less effective systems for damage repair.

Eros

It is difficult not to mention the correspondence between sex and aging and the psychoanalytic theories of Sigmund Freud, especially since Freud sought such a connection in his own writings (for example, in *Beyond the Pleasure Principle*). Freud theorized that humans are endowed with two major drives: the will to live and give life—the Eros, or "life instinct," as made manifest in sexual drives—and the "death instinct." At times he viewed the death instinct as independent of the life instinct and at other times dependent on the life instinct. In any event, what concerns us here is that Freud sought to find in biological theories of sex and death a foundation for these instincts.

Freud understood from Weismann that death is endogenous to the organism and took this as support for the existence of a death instinct. Undoubtedly, he would have been even more interested to learn that organisms are programmed to die. The implication in Freud's writings seems to be that the psyche in some sense knows that the body must die and so the psyche wants to die also. To put it simply: the psyche wants what the body must have—to die. It is well outside our concerns to consider the implications for the psyche of the fact that organisms are genetically programmed to die. However, there is one matter that should be appreciated in any such attempt: the genetic interests of the ego do not simply reside with the ego's own body. Other bodies living at the same time, especially collateral relatives, contain copies of the ego's genes and represent vehicles for their perpetuation. It would be an oversimplification of the genetic situation to say that an organism's biological interests end where its body ends and the outside world begins. It would be a surprise, indeed, if the psyche was not aware of the fact that its genetic interests may lie elsewhere.

More relative to our concerns is our theory of sex and what implications can be drawn for human behavior. As the quotation at the beginning of Chapter 8 shows, Freud found in Plato's theory of love (as told by Aristophanes) precisely the theory of sex he wanted—the purpose being rejuvenation and returning to a prior (error-free) state of things. Indeed, this is the correct view, if you accept the arguments I give elsewhere in this book. Freud interpreted Plato's theory as providing a biological basis for the "life instinct," the Eros, "the preserver of all things." For Freud, Eros was the most important of all psychological drives, all of which act to restore previous states. Love, according to Plato, led to a previous state of affairs "in which the losses caused by age are repaired by new acquisitions of a similar kind." Freud worried that future biological theories of sex might cast doubt on the connections he saw. It appears that just the opposite has occurred; if the ideas proposed here are correct, sex indeed originated as a rejuvenating force and tends to maintain the well-being of DNA and the immortality of life.

Sex and Immortality

In the simple protist *Paramecium tetraurelia* asexual clonal reproduction occurs by a mitosis-like mechanism. Under certain conditions, two cells fuse and have meiotic-like sex before splitting for a renewed cycle of asexual replication. Without sex, the replication rate slows down; eventually, after about 200 cell divisions, the clonal cell line dies out.

Sex is an effective way of coping with both genetic damage and mutations, as we will discuss in Chapters 6, 7 and 8. Since genetic error is the likely source of aging in the soma and since sex helps cope with genetic error, we can see why sex is necessary for immortality of the germ line. Clones age because they suffer, over time, an unendurable amount of genetic error, both damage and mutations. Asexual cells can make numerous repairs, but eventually the sheer accumulation of damage gets the better of them. In addition, Muller's ratchet (discussed in the next chapter) continually operates, resulting in an increasing load of mutations in asexual lineages. Germ cells, on the other hand, are created sexually, with recombination, so that their DNA is in perfect, or nearly perfect, order, and they are passed on to the next generation before too much else can go wrong.

Thus the germ line constitutes the only hold that any earthly creature has on immortality. "It is in this way that everything mortal is preserved," said Diotima in *Plato's Symposium*, "not by remaining for ever the same, which is the prerogative of divinity, but by undergoing a process in which the losses caused by age are repaired by new acquisitions of a similar kind . . . ; it is in order to secure immortality that each individual is haunted by this eager desire and love."

The Undoing of Sex

I hope to have shown that the theory of natural selection is by no means incompatible with the theory of 'the continuity of the germ-plasm;' and, further, that if we accept this latter theory, sexual reproduction appears in an entirely new light. . . . The object of this process is to create those individual differences which form the material out of which natural selection produces new species.

—*August Weismann*
Essays upon Heredity and Kindred Biological Problems

The idea that there is something special about eggs and sperms goes back to the nineteenth-century biologist August Weismann, who proposed that only the germ cells (gamete-producing cells) may pass on genetic information to offspring. As we have learned, our bodies are made up of somatic cells; they are a dead end in terms of propagating genes. There was one consequence of his idea that Weismann didn't like, however: it seemed to reduce the power of natural selection, and he was a firm believer in Darwin's and Wallace's new theory. To understand Weismann's worry, as well as his radical solution, we need to digress and consider Darwin's thinking at the time he proposed the theory of natural selection.

Weismann's Worry

Natural selection results, in part, from differences between organisms in their abilities to survive and reproduce, differences that are caused by the organisms' different traits. These differences must be heritable, that is, passed on from parent to offspring. If a male bird, by virtue of his beautiful song, mates more often than other birds, his sons will not enjoy the same success unless they too are endowed with their father's ability to sing. When Darwin proposed his theory, he was unclear on which differences between organisms were heritable. Indeed, along with the great French evolutionist, Lamarck, who preceded him, Darwin believed that acquired (nonhereditary) traits were inherited—traits such as the ability to play the violin, for example. Further, Darwin imagined that every slight difference between organisms that affected their ability to survive and reproduce was significant.

Weismann argued that differences acquired during a lifetime—differences, for example, between people in their ability to sing, or play the violin or write poetry—were not encoded in the genes and could not, therefore, be inherited. Acquired traits, Weismann argued, while possibly affecting the ability of the parent to survive and reproduce, would not likewise affect the offspring, and hence could not increase or decrease by virtue of natural selection.

Weismann worried that the noninheritance of acquired traits created problems for Darwin's theory since it restricted the pool of differences upon which natural selection could act. The large number of differences that arise during the lifetime of an individual in the somatic tissues (for example, an acquired limp due to a broken leg)—the "individual variability" emphasized some 40 years earlier by Charles Darwin—were not, in Weismann's view, available for natural selection, since they are not part of the "germ plasm" transmitted from parent to offspring.

Weismann set out to show that, in his words, "the theory of natural selection is by no means incompatible with the theory of 'the continuity of the germ-plasm' " by arguing that sexual reproduction provided an adequate level of germ-line variation to fuel evolution. By mixing the genetic differences already present in the population into ever new combinations, Weismann argued, sex would aid the population in adapting to new environments. He was aware of the basic problem with his suggestion: although sex may create new gene combinations in the production of offspring, these combinations are likely broken apart by sex when the offspring reproduce. Weismann's critics naturally seized on this problem. In his long and complex response, Weismann touched on many of the dominant themes existing today concerning the problem of sex: that sex has a twofold cost; that sex can destroy the new combinations of genes it creates; that fusion between unicellular organisms, such as bacteria and paramecia may provide nutrition; that sex is beneficial for the species; and, as the coming chapters will discuss, that sex serves to rejuvenate the germ line. To more rigorously discuss Weismann's proposal, we need some terminology.

Sex and Nonsex Genes

An organism's sexuality is, like all genetically based traits, determined by its genes. We will term these genes "sex genes"—or "asex genes" in the case of genes that cause an individual not to be sexual, that is to be asexual. Examples of sex genes would be genes encoding the recombination molecules, the meiotic apparatus, and the genes affecting how, when and with whom an organism mates. There are many other examples. We will often refer to both sex and asex genes collectively as sex genes, and differentiate between sex and asex sex genes when needed. The rest of the genes in the genome do not affect whether an individual is sexual or not. We term these genes "nonsex" genes, because they don't have anything to do with encoding sex (or asex) but rather affect other traits, such as height, eye color, intelligence and the like. We will primarily be interested in nonsex genes that affect the biological well-being of the organism—its fitness, or ability to survive and reproduce. But before we get to them we must remain with sex genes for a moment. Why are we talking about sex genes?

We want to understand why sex evolved. For any trait to evolve, there must be genes that determine the trait. This was one of Weismann's points, even though he didn't know about genes (acquired traits are not inherited because they are not in the genes, so to speak). So, for sex (or asexuality) to evolve, there must be genes that determine whether and how an organism has sex. These are sex genes.

The distinction between sex genes and nonsex genes is not as clean as I have just suggested, however. Recall from Chapter 3 how many male mating displays, the peacock's tail or deer's antlers, affect both their chances of mating and their survival. The genes involved are both sex and nonsex genes, but usually when we talk of "sex genes" we have in mind something more basic, such as whether an organism is sexual at all.

It is only rather recently that population geneticists have come to understand the principles that govern the dynamics of sex genes during evolution. Prior to this understanding, the problem of the evolution of sex was often approached through metaphors and vague analogies. In the pages that follow, we will first consider the metaphors and analogies that were initially used to understand the evolution of sex genes. Then we will recast these heuristics in terms of the underlying principles of population genetics recently discovered.

According to the "variation" point of view first advocated by Weismann, the benefit of sex genes lies in their producing new combinations of nonsex genes. Speaking very generally, we can imagine two kinds of nonsex genes: beneficial ones and deleterious ones (to the organisms that carry them). Accordingly, theories about the evolution of sex genes may be classified as either being concerned with beneficial combinations, or deleterious combinations, of nonsex genes. We'll consider deleterious combinations first.

Sex Retards the Ratchet in Small Populations

Recall how the early replicators eliminated errors by recombining them into "wastebasket" molecules (Figure 4, p. 23). This strategy would continue to operate once sex between cells was invented. We will take up meiosis in earnest in the next chapter, but you may know that chromosomes (one set from Mom and one set from Dad) break and rejoin during meiosis and that this can change the positions of genes on chromosomes. This is termed "crossing-over."

Consider two genes in a diploid organism and consider, for example, that one gene might affect seeing and the other might affect how fast the organism runs. To be explicit, we will refer to these genes by different letters and use lowercase letters for the mutations and capital letters for the corresponding nonmutated genes. Let us assume that the organism received one bad gene from each of its parents, so that the chromosomes have configuration *Ab* and *aB*. When this organism makes germ cells, crossing-over may move the mutations around, sometimes producing a chromosome that is entirely free of mutations (*AB*), sometimes producing one with two mutations (*ab*) and sometimes resulting in no change (*Ab* or *aB*). In this way crossing-over may produce chromosomes that are entirely free of mutations. If this were to happen with both germ cells, the offspring, by virtue of its having no mutations (being of genotype *AABB*), would be especially able to survive and reproduce and its good genes would come to predominate in the population and the bad genes (*a* and *b*) would be lost.

In this way, sex can help a population get rid of deleterious mutations. Of course, a population can never be completely free of mutations, since new ones are being created by DNA replication mistakes all the time. Still, sexual recombination can help keep their numbers in check. In contrast, in a population of organisms that reproduce asexually, without sex, there is nothing to keep deleterious mutations from accumulating. In 1964, the great American geneticist H. J. Muller used these ideas to argue that "an asexual population incorporates a kind of ratchet mechanism, such that it can never get to contain, in any of its lines [clones], a load of mutation smaller than that existing in its present least-loaded lines." Let's consider Muller's suggestion in more detail.

In any population, sexual or asexual, there will be individuals with different numbers of mutations, some having zero mutations, some one, two and so on. Initially, the class of individuals with zero mutations is what Muller meant by the "least-loaded line." There is always a chance that organisms die for no good reason—by chance, so to speak. Hurricanes or fires may destroy all life in a region, regardless of the features and numbers of mutations possessed by the organisms who happen to be there. Accidents happen to the best of us. True, the odds of survival may be better for the ones with fewer mutations, but in a world governed by chance even those with the best odds of winning can lose. Let us assume that all individuals with zero deleterious mutations fail to reproduce just by chance. At this point all individuals have at least one mutation and those who carry just one (let's say they weren't caught in the hurricane) represent the "least-loaded" line in Muller's quote above. The ratchet has clicked one stop.

The same thing could happen to these individuals—the organisms with just one mutation—as happened to the individuals with no mutations: they too might all die by chance, leaving just individuals with two deleterious mutations. When this happens, and it will happen eventually, the ratchet will have clicked two stops. In this way an asexual population will continually deteriorate by accumulating more and more deleterious mutations.

As mentioned, sexual parents can produce chromosomes—and offspring—with different numbers of mutations, some having more and some less than their parents. In a sexual population, offspring with no mutations can always be recreated, even if their kind had previously been lost by chance. For this reason, sex can prevent the ratchet from operating. In contrast, an asexual female passes on to her offspring all of her mutations and perhaps some new ones in addition. In an asexual population, there is no way of recreating individuals with no mutations. Once they are gone, the "ratchet" has clicked. And the ratchet will continue to click as the least-loaded individuals are lost by chance, until the population, burdened with its increasing load of mutations, eventually goes extinct.

The speed with which the ratchet operates in asexual populations, however, depends on many aspects of the species, especially the size of its populations (chance effects are more likely in smaller populations), the size of its genome (larger genomes have more mutations) and the severity of the mutations that occur (the ratchet operates more rapidly with mutations of small effect). The ratchet sets broad limits on the size of the genome that asexual organisms can attain. By avoiding the ratchet, a sexual population can keep a check on its numbers of mutations. However, the ratchet alone cannot provide us with a general explanation for the evolution of sex, primarily because it operates at the level of the whole population, and only in relatively small ones at that. In other words, while the ratchet may explain why asexual populations go extinct more often than sexual populations, it does not explain why asexual individuals without recombination don't continually arise and spread in sexual populations.

Interactions Between Mutations Can Favor Sex

Sexual parents produce offspring with differing numbers of bad genes; some will have many and some will have few deleterious mutations. As a result, sexually produced offspring differ in fitness; some do extremely well (those with few mutations), while others do not survive at all (those with many mutations). An asexual parent, on the other hand, produces offspring which all have the same intermediate number of mutations. In comparison to sexually produced offspring, asexually produced offspring do about average; none do extremely well, but none do extremely poorly either. Whether sexual parents or asexual parents do better (produce more surviving offspring) depends on how these mutations interact with one another in determining fitness.

In the extreme, we imagine a threshold number of deleterious genes, say four. Four or more mutations and you die. If an individual carries less than four deleterious mutations, the individual is just as well off as if it didn't carry any. The harmful effect of any one of the mutations is not expressed unless there are at least four of them present. This is what we mean by the genes interacting in some way—the effect of the fourth mutation is very different than the third. This threshold effect is an extreme form of gene interaction. It is not known to occur in nature, but it will help us think about the effects of gene interactions all the same.

With threshold selection, mutations accumulate in both sexual and asexual populations until the average number of mutations per individual is close to but not greater than the threshold (four, in our example). At this point the asexual population will be on the brink of extinction, because an asexual parent can not produce any offspring with less than four mutations (new mutations are occurring all the time). However, the sexual population and, more to the point, particular sexual parents, are able because of recombination and crossing over to produce offspring with less than four mutations. Sexual parents can also produce offspring with more than the threshold number of mutations, but these offspring do as poorly as the ones with exactly the threshold number. Because sexual parents can produce offspring with fewer mutations than the threshold, they have, on average, higher fitness, that is a higher number of surviving offspring, than asexual parents. This advantage to sex can occur for less extreme kinds of gene interaction than the case of threshold selection considered here, and is based (as was the ratchet) on sexual parents having greater variance than asexual parents in the number, or load, of mutations carried by their offspring.

"Now! Now!" Cried the Queen "Faster! Faster!"

We have been talking about good and bad genes as if this were some fixed property of the gene. While this may be true of certain genes (it is probably always bad to have a heart that doesn't work properly), for most genes, their suitability for reproductive success can change. Having bad eyesight may not be so bad if you are able to wear glasses, but it would be lethal if you were a hominid ape foraging on the African savanna. Whether genes are good or bad often depends upon the environment. The environment includes physical factors, such as temperature or moisture, but also includes other species who are themselves evolving. Predators, competitors and the like, are constantly changing in response to each other, leading to a kind of evolutionary arms race.

An evolutionary advance by one species—say, the production of a new toxic chemical by a plant—is experienced as a deterioration in the environment for the insect that feeds on the plant. Since the environment is always changing, a species must continually change so as to keep up. Without a continual supply of new beneficial traits, a species could not hope to keep up on such a treadmill. As the Queen says in Lewis Carroll's *Through the Looking Glass:*

> "A slow sort of country! Now, here, you see, it takes all the running you can do, to keep in the same place. If you want to get somewhere else, you must run at least twice as fast as that!"

Could sex help a population win this foot race? Weismann first pointed out that sex can bring together, in an offspring, beneficial traits that exist in different parents. He considered that this would benefit the species by providing a continual store of variability upon which natural selection could act. Subsequent work on this idea was done by some of the best minds in biology, the Englishman Ronald Fisher and the American H.T. Muller. We have already introduced Muller's ratchet and mentioned that Fisher discovered the runaway process of sexual selection. Fisher also wrote a classic text on evolutionary theory and invented many of the techniques of modern statistics, such as the analysis of variance and maximum likelihood. Muller did pioneering work on the nature of the gene. These workers followed Weismann's lead, and in 1932 they simultaneously suggested that sex would provide an advantage in the evolutionary race. After their rather brief comments and analysis, the problem of sex seemed solved—sex accelerates evolution, and that's a good thing. Case closed.

Science often debunks the obvious. The earth appears flat but we now know it isn't. Often in the practice of science, what seems obvious is not true. This has happened many times as evolutionary biologists have grappled with the question, Why sex? Sex appears to be for reproduction, but the sexuality of bacteria belies the limitation of this view. Sex appears to speed up evolution, but does it really? Sex appears to be for varying environments. But is it? We'll get to that in a moment. Now, let's see whether sex exists because it speeds up evolution.

The Vicar of Bray

Graham Bell, in his book *The Masterpiece of Nature: The Evolution and Genetics of Sexuality*, aptly characterizes Fisher's and Muller's view with reference to the Vicar of Bray, an English cleric who was noted for his ability to change his religion whenever a new monarch ascended the throne. This reference recognizes the core assumption: it is advantageous to easily adapt to changed circumstances. A sexual population is never committed to any particular composition and so can more easily change its genetic composition in the face of changes in the environment than a nonsexual population.

The Vicar of Bray hypothesis is the converse of the ratchet hypothesis. Most of the time, errors in DNA replication produce deleterious genes, but, rarely, very rarely, mistakes may improve the organism in some way. Have you ever kicked your TV to find that it works better? Even a flaky computer chip can be helped by a good bump! Random perturbations can help, sometimes. Instead of focusing on how sex might facilitate the removal of deleterious mutations, as does the ratchet and the threshold selection scheme discussed at the beginning of the chapter, the Vicar of Bray hypothesis focuses on how sex might facilitate the production of new combinations of *beneficial* genes. Consequently, the Vicar of Bray shares many of the same assumptions as the ratchet. However, because beneficial mutations are much rarer than deleterious mutations, this hypothesis has special problems that deleterious mutation theories do.

Progress in evolution ultimately depends upon the occurrence of mutations that are beneficial to the individuals that carry them. Fisher's and Muller's point was that, in an asexual population, each beneficial mutation so to speak, is on its own. Each must increase to completion before the next beneficial mutation can increase in frequency. If the two beneficial mutations existed together, they would compete with one another and only one would win. In sexual populations, however, different advantageous mutations can increase at the same time. Another way of saying this is, in asexual populations, beneficial mutations must increase one after the other, or serially, while in sexual populations they can increase at the same time in parallel. If you have ever tried to print from the serial port on your personal computer, you know that serial processing is much slower than parallel processing.

Let me be more specific. Consider what would happen in an asexual population, if two beneficial mutations arose at about the same time, in two different individuals. Let us assume that one of the mutations enhanced its carrier's ability to avoid predators by running fast, and the other mutation enhanced its carrier's ability to find food. After the mutations first arose, there would then be two kinds of individuals in the population: those more able to avoid predators and those more able to find food. (Since beneficial mutations are extremely rare, we ignore the possibility that they both occur in the same individual.) These two kinds of organisms will have different fitnesses, as it is inconceivable that two different organisms with different traits would have exactly the same fitness. Individuals more able to find food may find themselves on average better off than individuals more able to avoid predators. Or perhaps the opposite is the case. The important point is that the two kinds of organism will have different likelihoods of surviving to reproduce. As a consequence, the mutation with the highest fitness would win out and predominate in the population, while the beneficial mutation with lower fitness would be lost. The two beneficial mutations end up competing with one another and, although each is beneficial, in the end only one will be found.

Clearly, the best organism would be one that had both beneficial mutations (one that is both fast and able to find food). Such doubly fit individuals could then compete with individuals with no mutations and both beneficial mutations could increase together to completion. There are two ways in which the same organism might come to have both mutations. One of the mutations, say mutation *B*, might originate in an individual that already has the other mutation, say mutation *A*. But, as we have said, while the *A* mutation is rare, this is very unlikely.

Another way in which the same individual could get both beneficial mutations is by sex. Mating between two organisms, one who has *A* but not *B* and one who has *B* but not *A*, can produce offspring with both the *A* and *B* mutations. Again, while both mutations are rare, this mating is also unlikely. In addition, sex may undo the beneficial combination once it is created by producing offspring that again have one, or zero, mutations. At least with asexual reproduction, once an advantageous combination is achieved it will stay that way. For these reasons, it is not clear that sex is a more effective way of combining and keeping beneficial mutations together. And so it is not clear that Weismann, Fisher and Muller were right in assuming the obvious—that sex accelerates evolution.

Vicar of Bray Doesn't Work

About thirty years ago, a new generation of population geneticists started using mathematical models to study the Fisher-Muller process for the purpose of determining the conditions under which a sexual population would evolve faster than an asexual population. The results of this work can be understood in terms of one basic issue: do different beneficial mutations exist together in significant numbers in the same population. If they don't, sex cannot have an effect on the population's rate of evolution, because the chance is too small that mating might bring together two beneficial mutations in one offspring. The crucial factors determining whether different beneficial mutations exist together in the population are the size of the population, how favorable the mutations are (the intensity of selection), and the number of generations between the occurrence of favorable mutations (which depends on the rate at which favorable mutations occur). These factors are not independent.

In small populations mutations occur less often, since there is less DNA (fewer organisms) to mutate, so the time between beneficial mutations is longer. Furthermore, chance effects are more likely in small populations, so when a beneficial mutation occurs, it may be lost by chance. If the beneficial mutation is not lost by chance, it will take over the population in a relatively short time (because there are fewer organisms to be converted to the new type). These considerations all lead to the conclusion that in small populations different beneficial mutations are unlikely to be present in the same population at the same time. For this reason, sex has little effect on the rate of evolution in small populations. It is no better than asexual reproduction.

Just the opposite is true in large populations—different beneficial mutations are likely to be present in significant numbers in the same population, so the chance is high that two of them could be brought together by mating. Beneficial mutations occur more often in large populations (there are more organisms), and these mutations require longer periods of time to take over the population (there are more organisms to be converted). The intensity of selection affects the benefit of sex in a similar fashion. Slightly advantageous mutations take more time to transform the population and so spend more time at intermediate frequencies; during this time, new beneficial mutations may occur. So, the mathematical models show that sex accelerates evolution in large populations if most beneficial mutations are slightly advantageous to the organisms that carry them.

As pointed out by George Williams in his influential book *The Evolution of Sex*, these results reveal an internal contradiction in the Fisher-Muller theory. The theory works best in large populations, as we have just discovered. However, the theory also assumes population-level-selection: sexual populations should flourish, while asexual populations should be more likely to go extinct. This process of population-level selection, or "group selection" as it is more often called, is known to be most effective in small populations (I'll discuss why in a moment)—just the opposite of what is needed for the Vicar of Bray theory to work. This internal contradiction indicates the theory is deeply flawed. Fisher and Muller, two of the best minds in the development of modern biology, confronted the problem of sex and lost. What seemed obvious to them, that sex exists because it speeds up evolution, simply does not work the way they thought it should.

"Faster! Faster!" May not be a Good Thing

Putting aside for the time being the question of whether sex speeds up evolution, we might wonder whether it is always a good thing to speed up evolution. For the species, the bottom line in evolution is its persistence, that is, avoiding extinction.

There are two views on why species become extinct. In the first, the habitat of a species changes, and the species goes extinct because it can't adapt to the changes. For example, the temperature in a lake may increase due to changes in the atmosphere (perhaps as a consequence of global warming), and this may affect the hatching of eggs in a species of fish. To persist, the species would have to adjust its physiology through evolution to compensate for this change in temperature. The species' ability to do so will depend on its genetic variability, population size, mutation rate, whether it is sexual or not and other aspects of the population. If the species can't make the appropriate adjustments in its physiology, it will go extinct. According to this view, extinction results from internal constraints on the population (its size or mutation rate) and on the species' ability to adapt to its changing environment (whether it is sexual or not). For these reasons, both the Vicar of Bray and the ratchet should make extinction less likely in sexual populations.

George Williams promotes a second view of extinction: a species goes extinct because its habitat disappears entirely. Consider, for example, the remarkable species of green turtle that spends most of its life on the American shores but each year undertakes a journey of some 3,000 miles to tiny Ascension Island in the middle of the Atlantic Ocean to breed. This species is presumably well adapted to its habitat, but, even so, if Ascension Island eroded away, this well-adapted species would go extinct.

How does sex affect the likelihood of extinction if a species habitat disappears entirely? One might expect the greater rate of evolution of sexual species to perfect their way of life in an environment that will eventually disappear. Although sexual species may be better at keeping up with changing environments, they are never quite completely caught up. The constant reshuffling of genes keeps a sexual species from maintaining its adaptedness. In contrast, once an asexual species reaches the optimum phenotype, it has no difficulty maintaining it. A clone of dandelions that has evolved a high tolerance to acid soils has no difficulty staying that way.

We Can Explain Anything, or Nothing

How sex affects extinction depends on the view of extinction we accept. In the first view, asexual species are presumed to go extinct faster than sexual species because they can't keep up with an ever-changing habitat. This is the view that is implicit in the Fisher-Muller and Vicar of Bray views of sex and for this reason it is presumed to be advantageous to evolve as fast as possible.

The inability of a sexual species to precisely adapt itself may turn out to be a blessing in disguise if extinction results from habitats disappearing entirely. Due to its greater variability, a sexual species may consist of many different subpopulations that coexist in many different local habitats. For example, in one forest a local population of a bird species may be eating berries of one size and in another forest the birds may be using a slightly different kind and size of berry. As a result, the species may persist even if one of its habitats, say one species of berry just mentioned, disappears. On the other hand, an asexual species, because it can maintain a particular phenotype with precision, may become extremely well adapted to eating just one size of berry and go extinct if that particular kind of berry were to disappear.

However, there remains the issue of how fast the asexual species will evolve in the first place and whether it will ever find itself adapted with perfection to just one particular habitat. There is some evidence that current asexual species are just as variable as sexual species, although it is not clear if this is because they are recently derived from sexual ancestors or if they have accumulated variability through mutation.

In any event, our considerations of the effect of sex on the rate of evolution have led us to no certain conclusion at all. One line of thought suggests that sex may accelerate evolution, whereas another line of thought suggests that sex may retard evolution. Each of these mutually contradictory conclusions can be seen as being a good thing in certain situations and as a bad thing in other situations. It seems that we can explain the existence of sex no matter what its effects are. In other words, we have no explanation at all.

Group Selection for Sex: One Mutation, One Extinction

As if this weren't enough of a reason to look elsewhere for the answer to why sex exists, there are additional problems with the rate-of-evolution approach. As already mentioned, these hypotheses involve group selection: they explain why a group composed entirely of sexual individuals may evolve faster or slower than groups composed entirely of asexual individuals. Such concerns, while important in the long term, do not explain why, or even if, sex is advantageous to individual organisms (and their genes). Let me explain why this is a problem.

For the purpose of argument, let us give Weismann, Fisher and Muller what they want; we will assume that sex speeds up evolution, and that this is a good thing. Now, consider what happens to an asex gene (a mutation that makes an organism reproduce asexually) that arises within a sexual population. The new asexual individuals will have a similar level of genetic variability as that of their sexual counterparts, having recently been derived from them; yet they will not have to pay the costs of sex. The new asexual type need not allocate any energy or time to mating, and all their offspring have all their genes. The resulting twofold advantage will propel the asex genes into the sexual population, converting it to asexuality. This is individual selection in favor of asexuality and against the genes encoding sex. What can possibly keep asex genes from taking over and transforming the entire species to asexuality? According to the group selection explanation, it must be that the populations with more asexual genes evolve at a slower rate. But this will take a long period of time to have an effect, and during this time the asexual genes will continue to arise and increase.

Here is how group selection in favor of sex might work. The sexual species is, in reality, organized into local groups (like the local populations of berry-eating birds mentioned above). Just by chance, and possibly for other reasons, the asex gene will occur at different frequencies in different local groups. As we have noted, there is nothing to keep the asex gene from taking over a local population once the gene arises (ignoring any other effect that it may have, such as its lack of gene repair). Having just originated from the outcrossers, the asexual individuals are equally variable genetically. As a particular group comes to have a higher and higher frequency of the asex gene, the group may go extinct; it may be less able to adapt to changes in the local environment, it may suffer a build up of deleterious mutations (the ratchet), or it may overspecialize or underspecialize to its environment. The greater the proportion of individuals in the group that are asexual, the more likely is it that one of these events occurs. The groups with no or few asex genes should do the best and thereby spread their sex genes throughout the species.

According to this view, group selection between populations is favoring sex and individual selection within populations is favoring asexuality. Sex may benefit organisms and their genes for other reasons, perhaps because of damage repair or masking mutations, and these effects will help sex genes to more effectively compete with asex genes within the local populations. But that is another story, to be told in the coming chapters. At the moment, we are wondering whether the ideas of Weismann, Fisher and Muller are able to explain sex by themselves. In most situations, they're not.

A Crisis for the Theory of Evolution

We have found that every time an asex mutation arises within a sexual population, it will take over the population (it doesn't pay the costs of sex). Only extinction of the population can prevent asexuality from transforming the entire species. So we have arrived at a simple rule for group selection to keep a species sexual: for every asex mutation that occurs, the local population in which it occurs must go extinct (because the local population becomes asexual). Imagine a species of fish divided into several local populations occupying different stream basins in a mountain range. When an asex mutation arises, the local population will become asexual. This trend toward asexuality will cascade through the other populations until the entire species becomes asexual, unless each local population goes extinct when an asex gene arises within it. Local extinction is not an everyday occurrence, so either asexual mutations arise very infrequently or something is lacking in the group selection approach.

These problems with the group selection approach were first pointed out by John Maynard Smith and George Williams in the early 1970s and led to a renewed interest in the problem of the evolution of sex. For the previous 100 years, biologists had thought that the problem of sex had been solved by Weismann, Fisher and Muller. But all of a sudden the problem had no answer. The theory of evolution could not explain one of the most universal and ubiquitous features of the living world!

Balance Argument

Does sex benefit individual parents and their offspring or is it just good for the entire population or species? Even if we don't know what the benefit to individuals is, we can ask whether there is one. In his now classic book, George Williams said yes, we do know that sex is beneficial to individuals, even though we may not know why. In some sexual species—bacteria, viruses, aphids, daphnia (a small crustacean that lives in ponds and lakes) and some plants—both asexual and sexual kinds of organisms exist in the same population. These species are said to be "facultatively" sexual. If the two forms of organisms, sexual and asexual, weren't equally fit, one or the other should take over the population. This doesn't happen: both asexual and sexual forms are maintained. The sexuals appear to have made up for the costs of sex, because they have not been outcompeted by the asexuals. Consequently, so Williams reasoned, there must exist a short-term advantage of sex to individual organisms. Or does there?

There is a subtle catch in Williams's argument: the sexual and asexual forms must really coexist, they must live in the same habitat at the same time. The problem is that sexual and asexual forms are usually not ecologically equivalent, except in bacteria and viruses. In more complex facultatively sexual species, the sexual stage has adopted a specific function that is necessary for the species to maintain itself.

All complex organisms develop through different stages and these stages often have different purposes or are adapted to different environments. Consider the butterfly that was once a caterpillar, or the different castes of a honeybee hive: the workers, the queen, and the alates (winged form) all have different functions to perform in the interests of the hive. The fact that these different castes, or different life stages, are produced in a population doesn't lead us to wonder if one stage has greater fitness and if so why the other stages don't disappear. We don't wonder why worker honeybees don't outcompete the alate winged form, or why the caterpillar doesn't win out over the butterfly. We see that each stage or caste has a role to play in the functioning of the whole. And so it seems to be with the sexual and asexual forms of facultatively sexual creatures.

Consider, for example, daphnia, the common water flea. Sexually mated females produce a special "winter" egg that can survive if the pond dries up. When the pond refills with water, the winter eggs hatch and a population reestablishes itself with asexual females, who do fine as long as nothing happens to the pond. However, virgin birth cannot produce the special winter egg required to get through the winter (the usual time for ponds to dry up). A similar life history is practiced by some rotifers. Although in these species, sexual and asexual females may exist in the same pond, we cannot conclude that they have equal fitness and that the individual costs of sex have been paid.

Why is it that the special winter egg is produced only sexually; why can't an asexual female produce it? An advocate of the Fisher-Muller view might suggest that it was sexually produced combinations of genes that led to the capacity to make winter eggs in the first place. But these genes should now be present in asexual females, as these females were recently derived from winter eggs. So it should be possible for asexual females to produce a winter egg if it is advantageous for them to do so. A group selection advocate would argue that if the winter egg were produced asexually, it would have lower chances of surviving in next year's pond because it would not have the new combinations of genes required. It is difficult to rule out the group selection interpretation and put into practice Williams's balance argument that sex is advantageous to individuals. So we still don't have a compelling reason to believe that sex is beneficial to individuals except in bacteria and viruses.

Cost of Finding a Mate

The view that sex evolved to help a parent's offspring cope with changing and unpredictable environments is compelling, in spite of the problems mentioned above. As we have discussed in Williams's balance argument, some sexual species contain both sexual and asexual kinds of organisms. These facultatively sexual species often colonize discrete, localized habitats (like ponds, lakes, or host organisms), in which they reproduce asexually until they become numerous. When they are numerous, they disperse and colonize new but similar habitats (they move from pond to pond or to another host). In almost all cases, the organisms reproduce asexually within the local habitats, while the dispersing individuals are produced sexually. For example, many parasites reproduce asexually within or on their hosts but disperse to new hosts by sexual means. In grasses, an asexually produced runner forms a new individual near the parent, while the sexually produced seed is dispersed and settles far away from the parent. The bacterium *Bacillus subtilis*, discussed in the next chapter, becomes competent at sexual exchange when cells are numerous and immediately before production of spores, which are widely dispersed.

Correlations such as we have here between dispersal and sex cannot prove cause and effect. There is, for example, a correlation between the number of churches in a city and the number of jails—cities that have more churches also have more jails. Does this mean that there is a cause-effect relationship between churches and jails (could it be that going to church makes it more likely that you will commit a crime)? The reason for the correlation is that the numbers of churches and jails are correlated with a third factor, the size of the city, which acts as a common cause: larger cities have more churches and jails. A city with many churches will likely be populous and will likely have many jails.

Similarly, there is a third factor lurking in the data on facultative sex and dispersal. While it is the case that sex is correlated with dispersal, sex and dispersal are both correlated with a third factor— population density. According to an extensive analysis by Graham Bell, sex is elicited in populations in which the density of organisms is high. This makes perfect sense for the most elementary of reasons: why try to have sex if there are not others about to mate with? In dense populations it is easier to find a mate and the costs of mating are reduced. We will return to this point in later chapters.

The Lottery

Nevertheless, the pattern that asexual offspring are produced near the parent, while sexually produced offspring tend to be dispersed, suggested to George Williams that sex is advantageous in changing environments. Within the host, or near the parent, the environment is basically the same and the best strategy is for the parent to make exact copies of herself, since she has already demonstrated her ability to live there. In contrast, after the offspring disperse, the environment will likely be different and the best strategy will be to make a diverse group of offspring in the hope that one will find a new home. This seems to make sense.

Why is a diverse group of offspring the best bet? George Williams imagined that there would be a kind of lottery during the dispersal phase, when the offspring settle in a new and unpredictable environment. The parent's challenge is how to best produce an offspring with the winning ticket—the genotype most suited to the new environment. The answer seems obvious: produce different offspring to increase the chances of buying a winning ticket.

If you enter a lottery by purchasing several tickets, you are more likely to win a prize if these tickets have different numbers rather than if they all have the same number. This simple metaphor first provided by George Williams furnishes the basis for much current thinking on the possible advantage of sex to the individuals practicing it. The "prize" in the lottery of life and evolution is to have your offspring survive and pass on your genes. The lottery wheel of chance is the ever-changing and unpredictable environment. A sexual individual may purchase fewer tickets, that is have fewer female offspring, because of the cost of males, however at least these offspring are different from one another.

The basic underlying issue in the lottery metaphor concerns heritability in fitness. As we saw in earlier chapters, heritability in fitness is required for natural selection, and without it life is not possible. As long as the traits that make a parent fit to survive and reproduce are the same ones which make its offspring fit, there is no reason to produce a new combination of traits, no reason to reproduce sexually. The idea behind the lottery is that when environments change, so do the traits needed to survive and reproduce and for this reason it becomes advantageous to reproduce sexually.

Again, unfortunately, matters are not as simple as they seem. If offspring disperse a great distance, so that each offspring settles in a different environment, then the parents are entering a different lottery in each of the localities where their offspring settle. There is in this case, no advantage to being sexual. An asexual mother has tickets with the same number, but she is entering a hundred different lotteries. Holding different tickets for different lotteries is not any better than holding the same ticket for different lotteries. For the lottery metaphor to work, a parent's offspring must be entering the same lottery; this requires that offspring stay close to the parents and not disperse very far.

The lottery metaphor, no matter how intuitively appealing, clearly won't work as a general explanation for sex. We began by observing that dispersing offspring are often sexually produced. It seemed that sex must matter in new environments. We have found, however, that for sex to matter, dispersing offspring must not disperse too far. Another way of saying this is that brothers and sisters must end up competing with one another. Sibling competition occurs for some organisms (many of the seeds of an oak tree fall beneath the parent tree), but it does not occur as often as does sex. For the lottery metaphor to be apt, the new environment must be close at hand. But can the environment be so different if it's so close?

We have come to an impasse. What seemed so obvious, that sex is an adaptation to changing environments, has led once again to a simple contradiction. Twice we have begun with the obvious, or what appeared obvious, only to find that matters don't work out the way we expected. But evolutionary biologists are a clever group. They are not easily dismayed. True believers keep the faith even in the face of contradictory evidence. Armed with mathematical models, they have developed ever more elaborate and convoluted schemes to show what simple logic can't. They simply must find a way to show that sex is advantageous in changing environments.

Capricious Environments

In the past ten or so years a number of mathematical models have been developed in this quest. These models have returned both good and bad news for the believers. The good news is that sex is advantageous in changing environments. The bad news is that this only holds for environments that change in a certain way: the association between two states in the environment must flip-flop each generation. In other words, for sex to be advantageous in changing environments, it is not enough that the environment simply change, or that it be unpredictable; it is not even enough that the environment be completely random. For sex to be advantageous in a changing environment, the association between two relevant states of the environment must continually flip-flop each and every generation.

Environmental states must flip-flop each generation. Easy to say, but what does it mean? To understand what it means, we must be more explicit about the environment. This will lead us into a somewhat technical discussion of genetic associations between traits. You may find the whole matter rather arcane, but genetic associations is where the action is for these theories of sex. To understand Weismann's legacy in modern terms, we must learn something about genetic associations and the effect sex has on them.

Let's take the simplest situation possible and consider just two aspects of the environment—say, temperature and moisture—each of which can be present in either of two states—say, hot or cold and wet and dry. For a sex gene to outcompete an asex gene, the correlation between temperature and moisture, say, must change each generation. This means that if in one generation, hot environments tend also to be moist, and cold environments dry, then in the next generation the association must flip, so that hot environments tend to be dry and cold environments wet. For a sex gene to increase in frequency, so the mathematical models say, this flip-flopping must go on continually. As the eminent biologist John Maynard Smith said, "It is difficult to believe that God is as bloody-minded as that!"

Genetic Associations

We have been discussing flip-flopping of associations between the states of the environment, using temperature and moisture as examples. But what we really have in mind are two traits in an organism, each coded for by a gene (say, the A gene and the B gene) with different alleles, each subscripted by a number, (A_1, A_2 and B_1, B_2), that adapt the organism to the environment. So the required condition of flip-flopping of environmental states should really be translated into flip-flopping of genotypic states. As we have said, what is needed for sex to be advantageous is that the favored associations between the genes at the two loci must flip flop. For example, at one time the same numbered genes at the two loci should be favored (say, A_1B_1 and A_2B_2) and at a later time the mixed numbered genes should be favored (say A_1B_2 and A_2B_1). That way, sex (say, between $A_1A_1B_1B_1$ and $A_2A_2B_2B_2$ parents) can generate the newly favored set of associations when needed (say $A_1A_2B_1B_2$). In this example, we have assumed two environmental features each adapted to by a trait encoded by a single gene locus. In other examples that follow we consider a single trait controlled by two gene loci, instead of two traits each controlled by a single gene locus. The same requirement must be met for sex to be favored: the favored configurations of alleles at the two loci must flip-flop over time. (In the Chapter Notes we introduce the technical tool of "linkage disequilibrium" which is needed to describe these genetic associations more fully.)

Bloody-minded Parasites

Certain kinds of biotic interactions make the required flip-flopping easier to swallow. The British evolutionary biologist Bill Hamilton has argued that evolving interactions between parasites or prey should do the trick. Parasite-host interactions fit the Red Queen scenario mentioned at the beginning of this chapter. In the example of parasite-host coevolution, the environment of one species involves the adaptations of the other species, and vice versa. For example, the host species may evolve a new defensive chemical. This is experienced by the parasite as a change in its environment; to survive, the parasite population must evolve a means of detoxifying the chemical. If and when this happens, it is experienced by the host as a change in its environment. And so on, and so on. The evolutionary arms race between the two species can continue indefinitely, or at least until one of the species, no longer being able to adapt to the evolutionary changes in the other, goes extinct.

How might such a situation create the required flip-flop between the favored states for two genes involved in the interspecific interaction? One might conceive that genotypes A_1B_1 and A_2B_2 code for one type of host defensive chemical, say chemical I, and A_1B_2 and A_2B_1 for another, say, chemical II. Furthermore, letting lower case letters refer to genotypes in the parasite, genotypes a_1b_1 and a_2b_2 may detoxify the first host chemical and genotypes a_1b_2 and a_2b_1 the second chemical. We have constructed matters so the sex gene will increase. When most host individuals are producing chemical I, this will select for those genotypes in the parasite that can detoxify it. The increase of these genotypes in the parasite will be experienced as a kind of flip-flop since now host genotypes that produce chemical II will be at an advantage. These genotypes could not be produced in an asexual population that had fixed on chemical I but these genotypes could be produced in a sexual population. If and when the genotypes that produce chemical *II* become common, the parasite population will change its genetic constitution to the genotypes that can detoxify chemical *II*. Although one may find assumptions of the parasite model a bit contrived (after all it assumes what is necessary to get the sex gene to evolve) one can at least see how the required flip-flopping might occur. Again, the words of Maynard Smith: "It is not God, but a parasite, that is being bloody-minded." (Another way in which the required flip-flop might occur is discussed in the Chapter Notes.)

Sex Only Mixes

Sex cannot create a new gene: only mutation can do that. Sex simply assorts and recombines prior existing gene differences. If the population is already completely mixed, further mixing can have no effect and the busywork of sex is without consequence (from the point of view of creating new gene combinations). It's like stirring a bucket of paint. When the pigment has settled, the first few stirs have a big effect, but as the pigment and solvent become mixed, continued stirring has less and less of an effect. Once the pigment and solvent are completely mixed, further stirring has no effect. For stirring to have an effect, we must wait for the pigment to settle again. For similar reasons, sex can continue to have an effect on genetic variability only if there is some antagonistic process that keeps the genes unmixed.

The ideas discussed in this chapter can be understood as providing ecological or genetic situations in which it is a good thing to mix genes that are unmixed. To rigorously discuss these theories, we clearly need a measure of the amount of gene mixing existing in the population, or, equivalently, the degree of genetic associations present. In the Chapter Notes, we introduce the required tool, linkage disequilibrium, and discuss how the different theories of sex introduced in this chapter can be understood using this concept. Measures of linkage disequilibrium have been made repeatedly in the last twenty or so years, since the advent of molecular techniques like gel electrophoresis as a tool in studying genetic variation. Typically, the measured genetic associations are small. For the most part, naturally occurring populations of sexual organisms are already well mixed for randomly studied gene loci.

Sex Undoes What It Creates

Are we really to suppose that the significance of one of life's major adaptations is to be found in these small associations and second-order effects? As if the rarity of the cases in which sex produces something useful weren't enough, in the very next generation any fortuitous combination of genes produced by sex will itself be subjected to further random rearrangement by recombination and ultimate destruction. When Weismann originally proposed the idea, he labeled it "incredible" and devoted a long and involved discussion to justifying it for precisely this reason. It was apparent to him that sex could swallow up the very combinations of genes it created "because the deviations from a specific type occur in such rare cases that they cannot hold their ground against the large number of normal individuals." Weismann recognized that sex has two antagonistic effects on the level of population variability. Sex can create new combinations of genes, and, as discussed above, in certain circumstances these new combinations may benefit the organism in which they exist. However, the very sexual process that creates these fortuitous combinations of genes in one generation will break these combinations apart in the next generation. Sex undoes what it creates!

In Figure 1 the effects of recombination (crossing-over) between two genes with two alleles (A,a and B,b) are shown. Recombination (with probability r) can have an effect only in organisms that are heterozygous for both genes ($AaBb$). There are two kinds of doubly heterozygous organisms possible, those with AB and ab chromosomes and those with Ab and aB chromosomes. Recombination in an organism with chromosomes AB and ab produces Ab and aB chromosomes. Recombination in an organism with chromosomes Ab and aB produces chromosomes with AB and ab. Sex is always undoing what it is creating.

As a result of this undoing, a particular kind of flip-flop must occur in the favored gene combinations for sex genes to evolve by this means. The constant creation and destruction of new combinations of genes is precisely what is needed to keep up with an environment

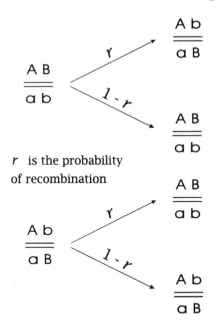

Figure 1. Sex undoes what it creates

that flip-flops each generation. What is normally a problem for this view is turned to an advantage in capricious environments, such as possibly host-parasite interactions. Whether these ideas have the generality needed to explain a ubiquitous feature of life like sex is currently an open question. I am doubtful, though. The little experimental evidence that exists is ambiguous, and the mathematical theory of host-parasite interactions and sex is poorly understood except in artificial situations. In addition, I don't see how this hypothesis could explain the molecular and cellular details of recombination that we will discover in the next chapter.

The constant undoing of gene combinations is not a problem for the deleterious mutation theories discussed at the beginning of the chapter. The new combinations of deleterious mutations created by recombination are removed by natural selection *before* they can be broken apart by recombination. After all, creating chromosomes especially laden with deleterious genes and quickly eliminating them is the whole point of these mutational theories. We will return to these mutation-based theories in later chapters, as they have a role to play in the theory we will develop.

Hopeful Theories of Sex

Since the time of Weismann, biologists have believed that the benefit of sex was to be found in the future, in the hope that the new combinations of genes created might find a use in new and unpredictable environments. Today, this hope remains the dominant idea, enshrined in college biology texts and popular references of one form or another. Unfortunately, there is little evidence supporting this view and, as we have found, many good reasons to question it.

Hopeful theories are intuitively attractive to the evolutionary biologist: after all evolutionary progress requires useful variations, and sex does, though rarely, produce them. All diversity in the living world comes from mutation and recombination. Nevertheless, there is little evidence supporting the view that sex evolved because of the diversity it creates. If the theories presented in this chapter have shown one thing it is this. In spite of the intense effort of some of the most gifted minds in biology over a period of more than sixty years, we still can't see in a clear and convincing way how sex can be selected for within a population on the basis of the genetic variability it produces.

Although hopeful theories have difficulty explaining why asexual mutants do not take over sexual populations, these theories do explain why the long-term prospects of asexual species are poor. Sex is a creative force. Asexual species do tend to proliferate less and to go extinct more quickly than sexual species. This much is clear. These long-term consequences of sex probably stem from its effect on genetic variability. However, to understand the evolution of sex we need to know why sex benefits individuals and the genes they pass off to offspring. Furthermore, the long-term effects of sex cannot explain the origin of sex within an asexual population.

The logical basis for the hopeful view has been explored many times. Using mathematical models, conditions have been found under which sex will evolve in changing environments. Unfortunately, the required flip-flopping limits the relevance of these models to special situations. It does not have the generality one would expect given the commonness of sex in the biota. Clever new theories are undoubtedly being invented as you read this book. What John Maynard Smith, a world authority on the evolution of sex, said over twenty years ago about the hopeful theories applies with equal force today: "I fear the reader may find these models insubstantial and unsatisfactory. But they are the best we have." There is a different approach, however, to the problem of why sex.

Twice Nothing

*I would rehearse a tale of love which I heard from Dio-
tima of Mantineia, a woman wise in this and in many
other kinds of knowledge, . . . She was my instructress in
the art of love, and I shall repeat to you what she said
to me, . . . "I mean to say, that all men are bringing to
the birth in their bodies and in their souls. There is a
certain age at which human nature is desirous of procreation—procreation which must be in beauty and not in
deformity; and this procreation is the union of man and
woman, and is a divine thing; for conception and generation are an immortal principle in the mortal creature, and
in the inharmonious they can never be. But the deformed is
always inharmonious with the divine, and the beautiful
harmonious. Beauty, then, is the destiny or goddess of
parturition who presides at birth, and therefore, when
approaching beauty, the conceiving power is propitious,
and diffusive, and benign, and begets and bears fruit: at
the sight of ugliness she frowns and contracts and has a
sense of pain, and turns away, and shrivels up, and not
without a pang refrains from conception. And this is the
reason why, when the hour of conception arrives, and the
teeming nature is full, there is such a flutter and ecstasy
about beauty whose approach is the alleviation of the pain
of travail. For love, Socrates, is not, as you imagine, the
love of the beautiful only." "What then?" "The love of
generation and of birth in beauty." "Yes," I said. "Yes,
indeed," she replied. "But why of generation?" "Because*

to the mortal creature, generation is a sort of eternity and immortality," she replied; "and if, as has been already admitted, love is of the everlasting possession of the good, all men will necessarily desire immortality together with good: Wherefore love is of immortality."
—Diotima instructing Socrates on the art of love,
in Plato's Symposium

In the beginning, according to Aristophanes speaking in Plato's *Symposium*, the world was populated by extraordinary human beings, each with four legs, four arms, two faces, two hearts—in short, double the number of each feature possessed by modern humans. Zeus, wishing to reduce the power of humankind, had the primeval beings split in two. The god Apollo sliced them down the middle, gathered up the cut skin, and smoothed out all the wrinkles, except a few at the navel. On Zeus's orders, he also turned each of their faces toward the cut side—the navel side—so that they might remember their division. Ever since, humans have lived in constant need to reunite with their other halves, and thus Aristophanes explained, man is motivated by love, "the love which restores us to our ancient state by attempting to weld two beings into one and to heal the wounds which humanity suffered."

Plato's Theory of Repair

Later speakers in the *Symposium* add a layer of complexity to Aristophanes' notion. Yes, love constitutes a desire for wholeness, the prophet Diotima tells Socrates, but it can exist only if the whole is itself something good. "Its object," she says, "is to procreate and bring forth in beauty . . . because procreation is the nearest thing to perpetuity and immortality that a mortal being can attain." Humans do not remain forever the same, as do gods, Diotima explains, but through procreation, "the losses caused by age are repaired."

The discussion in the *Symposium* is ultimately directed toward love's importance and complexity as a human emotion. The speakers draw a continuum from the physical love of two individuals to people's inherent love of wisdom. But in so doing, the ancient philosophers touch upon the solution to one of the most basic, unsolved mysteries in biology: Why is there sex?

Sex Makes Babies

Sex between mature adults produces babies that are youthful. The contrast between a baby's youthfulness and its parents maturity is as familiar as it is profound. The regenerative and rejuvenating powers of reproduction have appeared time and again in psychology, philosophy, literature and art. Plato's *Symposium* is but one example. Freud recognized the sexual instinct, Eros, as the preserver of all things. In *Germinal, Le Docteur Pascal* and *Fécondité*, the nineteenth-century French author Émile Zola develops the principle of fecundity into a theme of regenerative optimism. Many other examples exist in science, literature and art.

By the late nineteenth century, the idea that sex rejuvenates life was being discussed among scientists. However, there were problems with this view—the most serious of which was that scientists didn't know *how* sex could rejuvenate life. As a consequence, the hypothesis that sex rejuvenates life seemed mystical. In 1886, August Weismann simply dismissed the rejuvenation view by stating categorically that "twice nothing cannot make one." In his words,

> The whole conception of rejuvenescence, although very ingenious, has something uncertain about it, and can hardly be brought into accordance with the usual conception of life as based upon physical and mechanical forces. How can anyone imagine that an Infusorian, which by continued division had lost its power of reproduction, could regain this power by forming a new individual, after fusion with another Infusorian, which had similarly become incapable of division? Twice nothing cannot make one.

From our more modern vantage point, armed with our understanding of the workings of DNA, we can see a way around this apparent contradiction that Weismann saw as fatal to the rejuvenation view. The immortality of life must in some way be connected to DNA. DNA, like any physical thing, gets damaged and the resulting errors would destroy life unless there was some efficient means to get rid of them. DNA damage can be repaired by recombination if spare parts, in the form of genetic redundancy, are available. We have discussed in Chapter 2 how random exchange of segments between two damaged DNA molecules can result in a repaired molecule. As life continued to evolve, recombination got better at repairing DNA. Master molecular mechanics (so called recombinational repair enzymes) would evolve to make the process more precise and effective. The DNA passed on by parents becomes undamaged or whole again because of recombinational repair. Each damaged DNA molecule is, in Aristophanes' words, a "broken tally . . . in search of his corresponding tally," so as to become whole again. Two damaged complements can make one!

Two Kinds of Error: Mutations and Damage

We have mentioned briefly in Chapter 2 the two kinds of errors which occur in the transfer of genetic information. The first kind, a mutation, involves the change of one or more characters in a string of characters to other characters taken from the same alphabet. In the case of DNA, mutations involve the substitution, insertion, deletion or rearrangement of the four standard nucleotides, *A, T, G* and *C*. In the case of the English language, a mutation would correspond to the substitution of one Roman character for another or the deletion or rearrangement of Roman characters. Consider, for example, the word *message*. If, in copying or sending this word, the first *e* were changed to an *a*, a new word, *massage,* would be created. This is an example of a mutation, since a character has been changed to another character from the same alphabet. (In the example just given, both strings of characters, *message* and *massage*, were bona fide English words. This is incidental to the point that all characters in the two strings are valid Roman characters from the same alphabet.)

	Mutation	**Damage**
source string	message	message
received string	massage	m✿ssage

Figure 1. Mutation and damage

The second kind of error, a damage, involves the change of one or more characters to something that is not from the alphabet. In the case of the English language, a damage might correspond to some randomly drawn squiggle not belonging to the Roman alphabet. For example, changing *message* to *m✿ssage* would be an example of a damage, since a ✿ is not from the Roman alphabet. A damaged DNA string contains elements that are not one of the four nucleotides. Examples of DNA damage include physical cross-links between the two complementary strands in a DNA molecule, chemically modified nucleotides, and physical breaks in the strands of the DNA molecule. Damage interferes with DNA replication and transcription (a stage in the process of making proteins in which the information contained in DNA is expressed).

Because damage is not from the *A, T, G, C* alphabet used to encode the information, it can be directly recognized as erroneous by enzymes, whereas this is not possible in the case of mutations. Given the two strings *message* and *massage* above, we cannot say which is correct by simply comparing them. If we knew which string came from the source, we could compare the received string with the source and correct the received string if the two didn't match. Many error-control protocols used in sending computer bits work in this way. During DNA replication, there is a short period of time after a nucleotide is copied during which the new and old strings can be distinguished by enzymes and the mutation corrected. If the mutation is not noticed during this brief time it can no longer be recognized and corrected in the DNA molecule.

One might think that invalid words could be recognized—nonmeaningful strings of valid characters such as *messtge* or *fessage*. However, in the case of the DNA language, invalid words cannot be recognized either. Even in a human language, this is not always possible, since the meaning of a string of characters is a matter of convention. After all, we could agree that *fessage* means something. Meaningful strings of characters have no required sequence. Likewise, functional genes (analogous to valid words) have no recognizable sequence of nucleotides. Furthermore, there are no dictionaries providing character sequences of functional genes as there are for mean-

Figure 2. Single-strand damage

ingful words in human languages. There is simply no way of recognizing and repairing a mutation at the level of the DNA molecule. However, damaged strings, such as *me✿ssage* or cross-linked strands of DNA, can be recognized directly because they contain something that is not from the alphabet.

Because there are two strands in each DNA molecule, there is single-strand and double-strand damage. In the case of single-strand damage, only one strand is damaged and the other strand contains good information at the corresponding site. Both strands might be damaged, as in Figure 2, but at different sites. Sometimes (as in Figures 2 and 3) the characters of the DNA alphabet, the nucleotide bases, become modified so that they no longer can serve as a template for copying. Double-strand damage is more serious and is usually fatal to the cell. For example, sometimes the two DNA strands break or the two strands get linked to one another at the same site. Both these situations block DNA replication. In Figure 3, a double-strand damage is indicated by the distorted bases at the same nucleotide site in the two strands.

All living cells are endowed with a battery of repair enzymes whose only function is to fix DNA. Evolution would not have so equipped organisms if gene damage was not a serious threat to fitness. Environmental agents such as ultraviolet radiation, X rays

Figure 3. Double-strand damage

and chemicals damage DNA. However, much damage is endogenous—it originates from natural processes occurring inside the cell, such as metabolism.

Endogenous Damage

It is ironic that the very metabolic processes that sustain life are responsible for much of the damage that happens to genes. The food we eat is broken down into compounds like glucose and other carbohydrates. Organisms breathe air to extract energy from these compounds. Energy is extracted from carbohydrates in the form of excited electrons that travel along a series of carrier molecules in the cell, the electron transport chain, where they produce energy in the form of an energy-rich molecule called adenosine triphosphate, or ATP. It is these electrons that eventually combine with oxygen from air to form water. These energy-charged reactions are the furnace of the cell and ultimately keep the organism alive, but in the process the electrons may become reckless and wreak havoc in the cell. Highly reactive "oxidative" by-products are produced, such as the superoxide radical (O_2^-) and the hydroxyl radical ($\cdot OH$) produced from hydrogen peroxide (H_2O_2); these compounds damage DNA by chemically modifying the nucleotide bases, or by inserting physical cross-links between the two strands of a double helix, or by breaking both strands of the DNA duplex altogether. (The DNA-damaging properties of hydrogen peroxide make it useful in cleansing our wounds, where it mainly attacks the DNA of bacteria and parasites.)

Nucleotide bases are the characters of the DNA alphabet. When they become chemically modified they can no longer encode information or be copied; they must be removed by repair enzymes. The modified bases produced by oxidative damage are removed from DNA by excision repair enzymes and excreted from the body in urine where they can be studied. Thymine glycol and hydroxymethyluracil are just two of the many kinds of chemically modified bases that may be found in damaged nucleotides. Repair enzymes remove about 320 thymine glycol nucleotides from the typical human cell every day, and a comparable number of hydroxymethyluracil nucleotides are also removed. If we assume that at least 600 modified bases occur in every human cell every day (a low number), an additional ten double-strand errors should have also occurred, since it is known that modified bases and double-strand damage occur in a ratio of about 60 to 1. If we assume that only 1% of oxidative damage occurs in the nucleus where the cell's chromosomes reside (the rest occur elsewhere in the cell, as in the mitochondria), about one double-strand damage occurs in every cell in our bodies every ten days. This damage would substantially reduce fitness if unrepaired, since just one double-strand damage can block replication and cause cell death.

Since one double-strand damage is enough to kill a cell, every double-strand damage must be repaired for the cell to live. This is especially critical if the cell under consideration is that single cell, the egg or the sperm, that is passed on to an offspring. As we will see, recombinational repair between two DNA molecules during meiosis (the special cell cycle in which eggs and sperms are made), or during other forms of sex such as bacterial transformation, efficiently repairs these and other damage.

Coping with Error

Successfully coping with genetic errors is fundamental to life, for without the accurate transfer of genetic information from parent to offspring life would not be possible. As mentioned in Chapter 4, we can imagine four strategies that can be employed to deal with errors, and these strategies are similar, whether we are trying to keep DNA in working order or an automobile running.

In the first place, the cell should try to prevent damage from occurring. In the case of our car, we keep oil in the crankcase, vital fluids filled, and the tires inflated. Likewise, living systems have evolved defense systems to protect DNA from damaging agents: human skin, for example, may darken when exposed to the sun, to screen out ultraviolet radiation; cells themselves contain enzymes that neutralize DNA-damaging compounds; and the packaging of DNA within the nucleus of the cell probably was designed to shield the genes from caustic substances produced in the cytoplasm during metabolism. But despite all this protection, errors inevitably occur. There are three more strategies.

If our car is severely damaged, we may simply throw it away and buy a new one. Likewise, in multicellular organisms, cells with serious damage will not be able to replicate, while undamaged cells may divide to replace them. This is called cellular selection; its just like natural selection but occurs among cells in tissues of a multicellular organism. However, just as buying a new automobile requires money, so cell replication requires resources. In addition, damage levels must also be low enough so that there are undamaged cells capable of replicating to maintain a healthy population or tissue.

Another strategy to cope with errors is to hide or mask them. By keeping two or more copies of an essential component on board, so to speak, the backup can be used if one of the components becomes defective. The spare tire in our car's trunk is a familiar example. By swapping the spare for the flat, we avoid having a disabled car. The damage in the flat is not repaired, of course, as would become painfully clear if the flat stayed in the trunk and was forgotten until one of the remaining tires also became flat. The spare-tire strategy avoids the deleterious effects of errors not by fixing them but by hiding them and for this reason is only of limited use. This strategy of masking error does, however, play a role in our story, as we will see in the next chapter.

The final strategy of coping with errors is to repair them. In the case of a flat tire, we take it to the gas station for a patch. Spare parts for other kinds of automobile repairs are available to us at the local store. Where are there spare parts for repairing DNA? For DNA repair, a cell needs a backup source of undamaged and nonmutated genes. What could be better than using genes from another individual? True, this DNA will also have mutations and damage, but they will likely be at different positions in the DNA molecules, either in different genes or, if in the same gene, at different nucleotide sites.

For any important document, such as a will, I keep one copy at my office and one at home. If I spill wine on a page of the home will, I remove the damaged page and replace it with a copy of the corresponding page from the will at the office. Keeping two wills in separate locations is usually sufficient, so long as copies can be made, as it is unlikely that they would both become damaged at the same page at the same time. Likewise for DNA, a single backup copy of each chromosome should be sufficient. Even if both copies of DNA are damaged at several nucleotide sites, with some 10^{10} potential sites in the DNA of a complicated organism it is unlikely that these sites would be the same.

The four strategies enumerated above for coping with errors are all useful at different points in an organism's life cycle. However, when it comes time to give DNA to offspring, a parent cannot risk passing on any errors for such errors would severely disrupt the developing offspring. Having done everything possible to prevent mistakes from occurring, a parent gives a single cell to its offspring and must ensure that this cell is as free of errors as possible.

Redundancy

To repair its DNA, the cell needs access to spare DNA, just as an automobile mechanic needs spare parts. In cells, there is spare DNA wherever genetic redundancy occurs.

One form of redundancy exists within a single DNA molecule or chromosome; the two complementary strands of nucleotides provide identical information. If damage occurs in one strand, the error can be cut out by enzymes and the resulting gap filled with a chain of nucleotides copied off the other strand. This process, excision repair, occurs in most cells continuously.

A second form of redundancy—which is far more important in understanding the value of sex—exists within all diploid cells. As we know, in diploid cells chromosomes come in pairs. Each chromosome of a pair is redundant and, therefore, can effect repairs on the other. Even double-stranded damage (potentially lethal to the cell) can be fixed by recombination between the two chromosomes. I'll explain exactly how geneticists think this occurs in a moment. In the case of a double-strand damage, both complementary strands of a chromosome are damaged at the same, or nearby, nucleotide positions. The damaged portion is cut out, leaving a double-strand gap in the chromosome. Since there is no information left in this region, the information is obtained by recombination from the corresponding region of the second chromosome. Only recombination can repair double-strand damage: however, single-strand damage is also repaired quite efficiently by recombination. The process of recombinational repair can cause *crossing-over*—the movement of large regions of DNA from one chromosome to another—and this profoundly influences how genes are combined on chromosomes as we have discussed in previous chapters.

Recombinational Repair

We understand the fundamental role played by recombination and redundancy in repair of damage from numerous studies in a variety of organisms including yeasts, bacteria, viruses, fruit flies, and humans. Experimental studies show that when these organisms can't make recombinations, either because a second chromosome is not present in their cells or because they have defective recombination genes, they are extremely sensitive to agents that cause DNA damage, such as X rays or UV radiation. Furthermore, when these organisms are subjected to damage-causing agents, more recombinations occur in response. The recombination enzymes, those molecular mechanics who snip and zip DNA, are designed to make good repairs, not to create new gene combinations. In yeast, the mating, or outcrossing, aspects of sex also increase in response to DNA damage. These observations indicate that the primary purpose of sex is DNA repair.

Recombination is most advanced during meiosis, the special cell cycle that produces eggs or sperms, but it occurs in other kinds of sexual cells as well as in asexual cells. Whenever recombination occurs it appears to serve a role in DNA repair. Mitotic recombination in yeast cells, for example, increases dramatically when the cells are exposed to damage-causing agents. In humans, recombination between identical sister chromosomes increases in response to damaging agents (sister chromosomes are produced when a chromosome makes a copy of itself during mitosis). Although the frequency of recombination in mitotic cells is much lower than during meiosis (because of the absence of the special structures that promote chromosome pairing), the sheer number of mitotic cells means that many recombination events occur in somatic cells.

Recombination During Meiosis

Meiosis differs significantly from mitosis in two respects: the occurrence of recombination is much greater during meiosis and the daughter cells produced, the egg or sperm, are haploid instead of diploid (recall that haploid cells contain only one copy of each chromosome, while diploid cells contain two copies). When a haploid gamete produced by meiosis fuses with the gamete from its mate, outcrossing occurs and the resulting cell is diploid. The haploid-diploid cycle is a fundamental characteristic of all organisms that practice sex using meiotically produced gametes.

Meiosis is designed to ensure efficient recombination. Special structures exist for bringing the two chromosomes of a homologous pair together just so recombination can occur. The two members of each chromosome pair accurately align themselves next to each other, so that the genes on one chromosome are across from their counterparts on the other chromosome. (During mitosis, chromosome pairing does not routinely occur and so the frequency of recombination is far lower.) When the two chromosomes are finished recombining, each chromosome is not only free of damage but also contains a mixture of genes from each of the parents.

For the greater part of this century, scientists assumed that the primary purpose of meiotic recombination was to ensure that each germ cell contain a unique shuffling of its parent's genes. As summarized above, evidence accumulated to the contrary, however, and in the 1980s, geneticists discovered that the impetus for recombination is DNA damage. To initiate recombination, special enzymes make a double-strand break in one of the chromosomes of a homologous pair.

Risky Business

Imagine the risk involved in breaking a chromosome—it might not be put back together correctly. Indeed, recombination, when done incorrectly, is known to cause many different kinds of chromosomal aberrations that severely cripple their carriers. Why take such a risk? According to the Weismann legacy, it is so the offspring *might* receive new gene combinations; combinations that just *might* enable them to survive in new and unfamiliar environments. Does it really make sense for organisms to break their DNA on the off chance that a beneficial combination of genes just might be produced? I doubt it, especially since many offspring experience similar environments as did their parents and the parent is good evidence that the present gene combinations are working just fine. However, even if a better combination were created, it would be destroyed by the very process that created it when the offspring grew up and produced their own gametes. For the organism to take the risks inherent in breaking its chromosome, something must have been wrong with the chromosome in the first place. Probably there was damage already existing at the site of the double-strand break.

Based on extensive experiments, molecular biologists have developed a rather concrete picture of how recombination occurs between DNA molecules. This understanding does not support the view that recombination evolved to promote genetic variation, but it fits nicely with the view that the purpose of recombination is DNA repair. We will now consider the prevalent model of recombination during meiosis. Although the model is somewhat technical, I present it here, since it will help us gain a deeper understanding of the purpose of recombination. There are three general properties of meiotic recombination that should be kept in mind, if you get lost in the details. First, recombination begins with a double-strand break in one of the chromosomes. Second, the chromosome with the double-strand break initiates the recombination event and eventually receives information from the second chromosome. Third, recombination may or may not result in crossing-over between the large segments of the two chromosomes on either side of the break.

Double-strand Break Repair Model

Here is our present understanding of how meiotic recombination works in detail. Each recombination event involves a pair of chromosomes, which are drawn black and white in Figure 4. As we have already

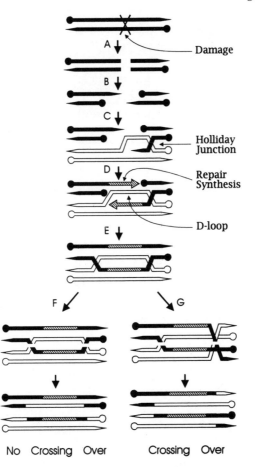

Figure 4. Meiotic recombination

mentioned, each chromosome is a double-stranded, or duplex, DNA molecule containing two complementary strands. The strands of DNA have a polarity, or direction, to them which is indicated in the figure by the arrowheads and circles at each end of each strand. The 3' ends are indicated by arrowheads and 5' ends by circles. The diagram magnifies the details of recombination within a very small region of the chromosome, equivalent to about one gene, with the vast majority of the chromosome extending to the left and right of the diagram off the pages of your book for about a half mile in each direction. The region shown magnified here in schematic form is where the chromosome breaks and rejoins with the other chromosome. The steps are as follows.

A. A double-strand break is made in one chromosome (the black DNA duplex in Figure 4). I am assuming that this is done at a site of damage as the data suggests.

B. An enzyme chews the DNA in the 5'-to-3' direction, converting the initial break into a double-strand gap and forming 3' single-stranded ends in the process. These free ends can invade the other DNA duplex.

C. One free 3' end invades the other homologous chromosome, (the white DNA duplex) forming a displacement loop (D-loop). The X-like structure is called a Holliday junction.

D. The D-loop is enlarged by elongation of the invading strand until the other 3'-ended strand can anneal with the D-loop from the other side. The striped arrow represents new DNA synthesis. Repair synthesis occurs on the top strand.

E. Repair synthesis and ligation fill in the gap in the upper strand. Elongation of the single-strand that initially invaded the D-loop results in enlargement of the D-loop and the growing D-loop eventually displaces the resident strand. The displaced strand migrates to the lower chromosome, forming a second Holliday junction. Degradation of some of the DNA may be required prior to joining the strands.

F and G. The Holliday junctions must be cut so the chromosomes can separate. The two Holliday junctions can exist in two equivalent "isomeric" forms. Thus, four configurations are possible at this stage, although only two are shown in the figure. These four forms are truly equivalent in chemical composition and cannot be distinguished from one another at the molecular level. Although there is no way of controlling which form a given Holliday junction takes, the outcome has profound consequences on genetic variability.

Consider the two alternatives shown after F and G in the Figure 4. In pathway F, the same structure is drawn as existed after step

E. When the two Holliday junctions are resolved by cutting and the loose ends joined, the ends of each chromosome have the same color, either black or white. There is no crossing-over. Only a few short patches of DNA have been moved around in the immediate vicinity of the double-strand break, but the vast majority of each chromosome remains unaffected by the recombination event (this outcome is called "gene conversion"—most recombinations during bacterial sex seem to be of this patchwork kind).

In pathway G, the left Holliday junction is as it is in F, but the right Holliday junction has been drawn in its other equivalent form. In F, the outside strands are straight and the middle strands have bends in them, and this is the case for both Holliday junctions. Which strands are drawn straight and which with bends is arbitrary. In pathway G, the middle two strands have been drawn straight for the right Holliday junction, and this forces the outside strands to be bent. When the Holiday junctions are cut and the loose ends cleaned up, one end of each DNA molecule is black and one end is white. In other words, there has been crossing-over. Crossing-over moves and recombines the thousands and thousands of genes that lie outside of the figure, to the left and right of the region of the break.

As mentioned, there are two other possible configurations of the chromosomes (four in all): one formed by making the inside strands at the left Holliday junction straight and the inside strands at the right junction bent (resolving this results in crossing-over), and one formed by making the inside strands at both junctions straight (resolving this results in no crossing-over). There are four possible outcomes: two produce crossing-over and two don't. If these outcomes were equally likely, there would be an approximately 50-50 chance of crossing-over when there is breakage and reunion of DNA, that is, when there is recombination. However, studies from yeast and other fungi show that crossing-over happens less often than that. Only about one-third of the time is there crossing-over when molecular recombination occurs. For this reason, it does not seem that meiosis is especially well-designed to promote crossing-over.

That is it for the molecular details of meiotic recombination. If you have not followed this technical detail, please keep in mind the following points. Double-strand damage initiates the process. When it is over, the damage has been repaired using information from the second chromosome. The process is not particularly effective at promoting crossing-over, as this happens less than half the time.

Sex and DNA repair is most advanced during meiosis, but it also occurs in bacteria and viruses. It is often easier to study DNA in simple creatures like bacteria and viruses, and, consequently, a great deal is known about recombination and sex in these organisms. Although it may seem like a tangent to you, we briefly consider the sex lives of these single-celled creatures. They must be having sex and making recombinations for the most fundamental of reasons. How and why do they do it?

Sex in Bacteria

Transformation is a naturally occurring process in which some species of bacteria bring DNA into the cell and recombine it. Although possessing the fundamental properties of recombination and outcrossing, transformation has several unique aspects when compared with meiotic sex. In meiotic sex, each mate through its egg and sperm provides one-half of its offspring's genes. In contrast, in bacterial sex there is a recipient and donor of genes. The donor DNA brought in by a recipient cell is released by neighboring bacteria. In meiosis, sex is connected with reproduction. We have already mentioned how this may mislead the casual observer into thinking that the purpose of sex is reproduction. In bacteria, reproduction is like mitosis and does not involve sex. For example, while the bacterium *Bacillus subtilis* is reproducing rapidly, it almost never has sex. Competence at sex occurs primarily when the cells have stopped reproducing and the population has saturated its environment. At this point, DNA is brought into the cell from outside and recombined.

Transformation appears to be a highly evolved and purposeful feature of these bacteria. The recipient goes to great expense, in terms of both time and energy, to accomplish it. First, the recipient cell must develop competence—the physiological state in which it can have sex. After a cell becomes competent, it binds donor DNA from its mate to specific receptors on its cell membrane and processes it for uptake. Once inside the recipient, the donor DNA is protected and recombined with the chromosome at homologous nucleotide sites. Although much is known about the details of the various stages depicted in Figure 5, their overall purpose is poorly understood.

The repair hypothesis predicts that transformation should function in DNA repair. Several aspects of sex in *B. subtilis* make sense when seen from this point of view. The same genes that control transformation also function in DNA repair. In addition, competence at sex is greatest when genetic redundancy within the bacterium is lowest. When bacteria are dividing rapidly, they usually have more than one chromosome per cell, since DNA replication often outpaces cell division. (This situation allows for the well-studied process of post-replication recombinational repair to occur between two sister chromosomes present in a dividing cell.) However, when conditions are not right for cell division—vital nutrients are not available or have been used up—then bacterial cells stop dividing. At this point, they

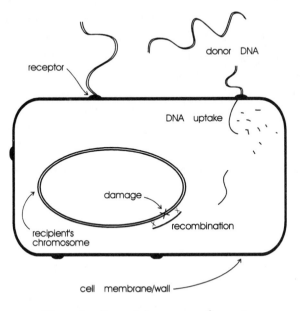

Figure 5. Bacterial sex: transformation

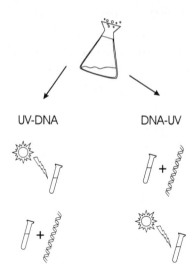

Figure 6. Experiments on bacterial sex

typically have just one copy of their chromosome and have no backup chromosome with which to make repairs. The needed spare parts are not present in the cell. Transformation probably functions during this period to provide the needed spare DNA molecules for use in gene repair. (For this reason, in Figure 5 I have assumed a damage in the recipient's chromosome at the site of a recombination event.)

In my laboratory we have conducted experiments using a naturally competent bacterium (*B. subtilis*) to study this idea. We have studied the survival of sexual and nonsexual cells in environments with differing amounts of a DNA-damaging agent, such as UV radiation (indicated by a sun in Figure 6). A population of cells is grown for several hours until the cells saturate their environment—a flask containing growth medium—at which time a portion of them, about 10%, become competent at transformation (the majority of the cells remain noncompetent). In one kind of experiment, termed a "UV-DNA" treatment, we damage the entire population with UV light and immediately afterwards give the recipients (indicated by a test tube in Figure 6) good donor DNA that has not seen UV light. In the reverse experiment, termed a "DNA-UV" treatment, we switch the order of UV and DNA treatment, that is, give the recipients donor DNA first and afterwards damage them with UV light. After the manipulations are complete, we count the numbers of transformed (sexual) and total cells (which are primarily asexual, as the majority of cells are not competent). The reason for comparing the results from these two kinds of experiments is that in a UV-DNA treatment the donor DNA could be used by the recipient to repair any damage it may have received from the UV light. This is not possible in a DNA-UV treatment, because the recipient is done having sex before it receives damage from the UV light. Therefore, the results of the DNA-UV experiments serve as a control (a point of reference) to which we can compare the results of the UV-DNA experiments.

The repair hypothesis predicts that the survival of sexual bacteria should be greater than nonsexual bacteria in UV-DNA experiments, because they can make repairs using the good donor DNA. This is what we observe in our experiments. In DNA-UV experiments, in which the cells have sex before they are subjected to UV, the two kinds of bacteria (sexual and nonsexual) have approximately the same survival rate. This, too, is what we expect, because having sex shouldn't be of any use in repairing damage that comes later. The higher survival we observe in UV-DNA experiments likely results from UV damage stimulating transformation. Bacterial cells are more likely to have sex if they are damaged in the first place.

In nature, the DNA of both partners will likely be damaged to the same extent, so we tried damaging the donor DNA before giving it to the recipients to see if this made any difference in the outcome of our experiments. It didn't: sexual cells still survived better than nonsexual cells in UV-DNA experiments. Why shouldn't it matter whether the donor DNA is damaged? You might think that by bringing damaged DNA into the cell, the bacterium might disrupt its otherwise healthy genes. However, recombination events during transformation appear to be targeted to damaged sites in the recipient's chromosome. With such a large number of possible sites in the DNA, the damage in the donor and recipient will likely be in different places in the DNA molecule. If recombination events are targeted to damaged sites in the recipient's chromosome (as shown in Figure 5), the needed fragment will likely be free of damage at the needed nucleotide site(s).

The bacteria we study are not fussy about whose DNA they bring into the cell. While they may bring other species' DNA into the cell, only DNA from their own species is recombined. We have used this property to make sure that the higher survival of transformants in the UV-DNA experiments depends on the information contained in the donor DNA molecule and not some other characteristic, such as its nutritional value. Recall, from Chapter 2, how some biologists think that cells first started having sex for nutritional reasons. We have considered this hypothesis by using another species DNA as donor material. This DNA is still brought into the recipient cell. Because it contains different sequences of nucleotides, different information, it should not be of any use to the recipient for gene repair. (For similar reasons, it wouldn't help to fix the will at my office by using a page from my tax return at home!) However, if the transforming DNA was being used as food, one species' DNA should be just as good as another (the chemical content of different DNAs are about the same). As expected by the gene repair hypothesis, sexual bacteria who were given donor DNA from another species did not show any increase in survival in a UV-DNA experiment.

These experiments are consistent with the view that transformation functions in DNA repair. However, other workers have shown that natural transformation also functions to create novel combinations of genes. It probably does both. The use of transforming DNA in gene repair can result in the transfer of genes if the donor's genes differ from the recipient's at the site of repair. So the two views, one that sex functions in gene repair and the other that sex functions to create novel combinations of genes, are not mutually exclusive. Indeed, genetic novelty is a consequence of gene repair. Ironically, in the quest to maintain genes, the gene combinations can change.

Multiplicity Reactivation

Even the simplest of all living creatures, viruses, practice a kind of sexual gene repair. Viruses are very simple creatures, so simple that some biologists don't even think of them as being alive. Viruses are little more than a few genes wrapped in a protein coat. They reproduce by infecting cells of bacteria, animals or plants, injecting their DNA (or RNA) and then diverting the metabolism of the host cell to making virus progeny. In the process, the host cell is killed. Viruses contain just one copy of each gene; they are haploid. Usually just a single virus infects a host cell, but sometimes two or more viruses infect the host at the same time. Such multiple infections provide for genetic redundancy for virus genes, just as diploidy does, and just as transformation does for bacterial cells. During multiple infections the genomes of the coinfecting viruses may recombine, resulting in virus sex (Figure 7).

Like all creatures, viruses are sensitive to agents that damage DNA—sunlight, X rays and the like. And when their DNA contains damage, they make fewer progeny, especially when they infect a host cell alone. After applying different doses of a DNA-damaging agent, like UV radiation, to the infecting viruses, the numbers of progeny produced by single and multiple viral infections can be studied. Scientists have found that when a damaged virus infects a cell alone, it produces far fewer progeny than when it is undamaged (and infects

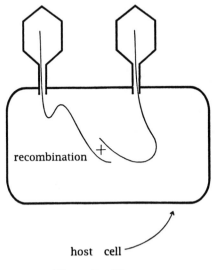

Figure 7. Virus sex

alone). However, when two damaged viruses have sex by infecting a host cell at the same time, they produce about as many progeny as when they are undamaged and infect alone. This benefit of sex in the face of gene damage has been shown to result from recombinational repair.

Viruses can control their own sexuality and they appear to do so as an adaptive response to DNA damage. In the bacterial virus phage T4, one or two genes control whether an infecting virus allows a second virus to coinfect the same bacterial cell. At low levels of DNA damage these genes prevent multiple infections, so that virus reproduction is asexual. The virus is said to be "immune" to multiple infections when these genes are expressed. But these immunity genes are turned off when the first virus that gets into the host cell is damaged. With the immunity genes turned off, a second infection by another virus—sex—becomes much more likely. Therefore, virus sex is more likely in damaging environments than in nondamaging environments, as predicted by the gene-repair view.

In Haploids, Outcrossing is Needed for Repair

These results show that bacterial and viral sex functions to repair DNA. Although experiments of this kind can never prove the hypothesis that sex first evolved for purpose of gene repair, they give us confidence in the idea. Recombination originated in organisms similar to present-day viruses and bacteria and there is growing consensus among experts that DNA repair was the reason. In these simple haploid creatures, outcrossing is needed to provide a second DNA molecule—the required spare part—for recombinational repair. Consequently, both aspects of sex, recombination and outcrossing, are needed for effective gene repair in single-celled haploid creatures.

Recombination is a process of nearly universal occurrence and functions to repair damaged genes whenever and wherever it occurs: during meiosis, during mitosis or during sexual exchange in bacteria and viruses. It is because of recombinational repair that Weismann was wrong—twice nothing *can* make one. The outcrossing aspect of sex is found less universally, but it is found in the simplest and most complex organisms on earth. As we know, outcrossing could have originated for DNA repair, however other ideas, such as nutrition or infectious transfer of selfish genetic elements, are also plausible. Complicated organisms like ourselves are made up of many diploid cells and have plenty of spare gene parts close at hand. For this reason in diploids outcrossing is no longer necessary for gene repair. In the next chapter we turn our attention back to these more complex diploid organisms. Why do they continue to outcross when having sex?

Chance and Necessity

Everything existing in the Universe is the fruit of chance and of necessity.

—*Democritus*

But once incorporated in the DNA structure, the accident—essentially unpredictable because always singular— will be mechanically and faithfully replicated and translated: that is to say, both multiplied and transposed into millions or billions of copies. Drawn out of the realm of pure chance, the accident enters into that of necessity, of the most implacable certainties. For natural selection operates at the macroscopic level, the level of organisms.

—*Jacques Monod*
Chance and Necessity

In biology, chance is a factor of profound significance. We have already mentioned how evolution opportunistically evaluates chance events, mutations and recombinations—that if life were to begin again on this planet, even given identical conditions, the living world would be different the second time around. Most species owe their existence to chance and most go extinct because of chance. The dinosaurs flourished for over a 100 million years only to be wiped out in an instant by, if a current theory is correct, an asteroid that happened to collide with earth. As it turned out, we were the big winners. Following that run-in between earth and the asteroid some 65 million years ago, the era of the reptiles ended and the age of mammals began. Our small ratlike ancestors would flourish in the aftermath of the dinosaurs' demise. It didn't have to turn out that way, for there wasn't anything that special about those ancient rats, our ancestors. As we know, matters could have turned out differently.

From Accidents to Design

In spite of, and for many reasons because of, the occurrence of chance events, life is possible. The relentless push of entropy has been delayed on the planet Earth. The second law of thermodynamics states that everything is getting more and more disordered, that entropy (disorder) is always on the increase in a closed system. The universe is a closed system; this law must be obeyed. Nevertheless, higher states of order are possible in local regions of the universe (such as on our planet), even though the whole system (the universe) is getting more and more disordered. Each living thing, although precariously close to its own extinction, is kept alive by food and warmth. Ultimately, it is the sun that, as it tends to its own death, fuels life on earth.

Evolution acting on chance events plays no small part in this process. The energy from the sun fuels the bacteria and plants that feed the living world. This energy is channeled eventually into offspring, some organisms producing more than others, according to the traits and genes they possess. To reproduce, an organism must pass on copies of its DNA. Accidents happen in copying DNA, and rarely—very rarely—the resulting mistake improves the offspring's chances of surviving to reproduce. These beneficial mutations fuel the evolutionary process by providing genetic variants upon which natural selection can act. Sex and recombination greatly assist the creative power of natural selection by providing new combinations of existing genes. Feeding on these mistakes and chance combinations, natural selection builds ever more complex forms in response to new challenges and by so doing abates biological decay.

For example, bacteria normally cannot grow in the presence of streptomycin, a drug that kills bacterial cells by binding to their ribosomes (needed structures in the cell for making proteins). In a population of about 100 million bacteria cells, about 1 cell will have undergone a random mutation somewhere in the gene that encodes a particular protein that makes the cell's ribosomes. This random change may affect the ribosomes in such a way that streptomycin no longer can bind to the ribosomes and kill the cell, so the mutation makes the cells resistant to streptomycin. In an environment in which streptomycin is present (for example, in a human being who is taking antibiotics), such a mutated cell would be at an advantage in competition with other bacterial cells that were sensitive to streptomycin. Although some mutations are beneficial, the vast majority are bad for the organisms that carry them.

Most Mutations are Bad

Mutations are random changes in the sequence of nucleotide characters in the DNA molecule. They are unavoidable. No process can be perfect and when DNA is replicated mistakes are made, usually at very low frequency. For example, instead of putting the preferred A across from a T, the replication enzyme may put in place a wrong character, say a C. Other kinds of mutations are possible—rearrangements of sequences, loss of characters, addition of characters, and the like. Most of these mutations have a deleterious effect on the organism for the simple reason that if you randomly change a complicated machine, such as an automobile engine, you will more likely decrease its performance than improve it. Consequently, most random changes to an organism's genome decrease its ability to survive and reproduce in its environment.

Bad genes are a fact of life. Each of us carries several extremely nasty mutations that would kill several times over if they were expressed. Fortunately, most of the time in our diploid cells these genes are not expressed. Because most mutations are recessive, they are masked by the corresponding good gene at the same locus in the other chromosome. In addition to these lethal genes, numerous less harmful mutations hide out in the chromosomes in the cells of our bodies, waiting to be expressed if they become homozygous (present in both gene copies in a diploid cell).

Why Is Outcrossing Needed in Diploids?

In the last chapter, we considered how sex may have originated for the purpose of damage repair in simple prokaryotes similar to present-day bacteria and viruses. Because bacteria and viruses are often haploid, containing only a single copy of each gene, gene repair of double-strand damage requires that a second copy of the damaged gene (the spare part) be brought into the cell from the outside by outcrossing. Consequently, in the case of haploids, the group of organisms in which sex almost certainly originated, both recombination and out-crossing (sex) are necessary for repair of damaged DNA.

As life continued to evolve and diversify, diploid organisms emerged as the dominant life form. Diploid organisms have two copies of every gene in each cell, and it is unlikely that each copy would become damaged at exactly the same site. As far as we know, damage occurs randomly in each DNA molecule, and the nucleotide site that gets damaged in one molecule is independent of the site that becomes damaged in another molecule, even if the molecules are present in the same cell. For this reason, there will usually be a good copy of a damaged gene present in a diploid cell. Consequently, it seems that outcrossing is no longer needed for gene repair. Why, then, has outcrossing been maintained in the majority of modern organisms who are diploid? This is the main question I wish to answer in this chapter.

Haploids to Diploids

A clue to answering this question lies in the likelihood that diploidy became the dominant stage in the first place. All organisms pass through both a diploid and a haploid stage, although the diploid stage may be very short-lived. For example, as we discussed in the last chapter, diploidy is a transient stage of gene repair during the sex lives of bacteria and viruses. But as organisms became more complex, the diploid phase became more pronounced. Our gametes, the sperm and egg cells, are the only remaining haploid cells humans have. Our bodies are made up almost entirely of diploid cells, and this is true of most of the organisms you are likely to be familiar with. Why did the haploid stage become reduced and the diploid stage become dominant?

Greater complexity has been one of the hallmarks of recent evolution. Plans for a skyscraper occupy many books, while plans for a one-room cabin can be drawn on the back of an envelope. For similar reasons, it requires more genes to build a complex organism than a simple one. More genes means more DNA, and for a fixed error rate per replicated nucleotide, this means more mutations. To make complex organisms, ways had to be found of first avoiding and then coping with the greater numbers of deleterious mutations that occur when copying more and more nucleotides.

At first this was accomplished by improving the care with which each nucleotide was replicated. A simple virus, like the flu virus, has just a few genes (about 8) requiring about 10,000 base pairs to encode. With such a small number of genes, a flu virus can tolerate a relatively high mutation rate of about 1 mistake in every 10 million nucleotides copied (mutations per nucleotide copied $= 10^{-7}$) and still produce mostly viable offspring. With the care the parent virus takes to replicate its 10,000 base pairs, about 1 out of every 1,000 progeny will contain a mutation somewhere in its genome, which is the entire set of genes needed to make the organism (10^4 nucleotides per genome $\times 10^{-7}$ mistakes per nucleotide $= 10^{-3}$ mistakes per genome). Natural selection will eliminate the less fit progeny (the ones with deleterious mutations), so the others can reproduce and maintain a healthy population. (In addition, the rarely beneficial mutation may still occur and improve the well-being of the population.)

Fidelity in DNA Replication

As more complex forms of single-celled organisms evolved—for example, more complex viruses, bacteria, yeasts and molds—more genes were required to encode the additional features, like the ability to use new resources, more complex internal cellular structures and flagella to move about the environment. These features required vast amounts of additional genetic information, about 4 million nucleotides for bacteria and 50 million nucleotides for a yeast cell. These additional nucleotides encode the information for about 5,000 to

10,000 genes, compared to the 8 genes present in the flu virus. To accommodate these additional genes and still produce mostly viable offspring, more care was taken in replicating nucleotides, so that the mutation rate per nucleotide copied dramatically decreased. Two mistakes are made in a bacterium, and less than one mistake is made in a yeast cell, for every 10 billion nucleotides copied! This is an error rate so low that bacteria and yeast have about the same mutation rate per genome (of about 1 mistake in every 1,000 offspring produced) and thus produce a similar proportion of mutant offspring, as does the far simpler flu virus with its 8 genes.

In Figure 1, the genome size (numbers of base pairs in the complete set of DNA needed to make the organism) and the mutation rate per nucleotide (chances of putting an incorrect nucleotide in a

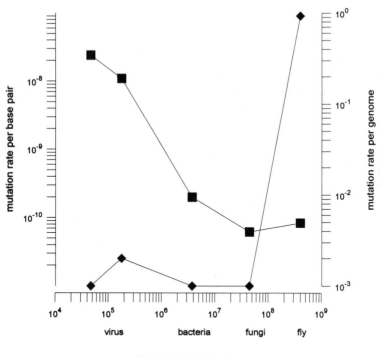

Figure 1. Mutation load and genome size

new string, for example, a T across from a *G* instead of a *C*) are given for a variety of organisms, from a simple virus to a complex insect. Also calculated for each organism (the product of the first two numbers) is the overall mutation rate—the chance of there being at least one mutation every time the genome is replicated.

As the size of the genome increases from about 10,000 to between 10 million and 100 million base pairs, the mutation rate per nucleotide drops from about 1 mistake in 100 million nucleotides copied to 1 mistake in 100 billion nucleotides copied, so as to maintain a roughly constant mutation rate per genome of 1 mistake in every 1,000 genomes copied. As we have mentioned, the problem of deleterious mutation was handled successfully in these organisms by increasing the care with which each nucleotide was replicated. However, for the common fruit fly, *Drosophila melanogaster*, the mutation rate per genome jumps three orders of magnitude to about 1 mutation in every offspring produced. Apparently this increase is unavoidable; the fly genome is much larger, there are more genome replications per fly generation (after all, the fly is made up of many cells while the other organisms in the figure contain just one cell), and there was no improvement in the fidelity of DNA replication (in the fly the mutation rate per nucleotide is still 1 mistake in 100 billion nucleotides copied, the same as in yeast). But why didn't evolution improve the care a fly takes when it replicates each nucleotide?

One error in 100 billion (10^{11}) nucleotides copied is a very small number, and to make it even smaller is possible, as recent experiments have shown, but only by drastically slowing down the rate of replication and/or by increasing the numbers of enzymes and genes devoted to this task. It appears that with an error rate of 10^{-11} evolution had reached a point of diminishing returns, where further increases in the fidelity of replication were simply too costly in terms of the time, energy and resources needed.

A far simpler, though short-sighted, solution was available—just stay diploid most of the time. The majority of mutations are recessive, or nearly so, and are masked in the diploid state. So it is no surprise that the fruit fly, *Drosophila*, is the only organism in the figure that is diploid. Diploidy likely became the dominant stage because of masking of recessive mutations. In the long term, nothing was to be gained by switching to diploidy, since recessive mutations accumulate in the genome and eventually begin expressing themselves as their numbers built up. And, with more DNA to replicate in a diploid cell, the mutation load soon exceeded that present before the switch to diploidy. This is where outcrossing and sex come in. Diploidy alone is not the answer to the problem of deleterious mutations; diploidy must be combined with outcrossing, that is, sex.

Before explaining this, let us review a few basic concepts from genetics. In diploid cells each gene is present in two copies. Any particular allele (or form of a gene) may be homozygous, that is, present in both copies, or heterozygous, that is, present in only one copy. If a gene is in the heterozygous state, the other chromosome carries a different allele of the same gene (alleles for blue and brown eye color, for example). In heterozygotes, the two alleles are different by definition. Whether alleles are heterozygous or homozygous is important for two reasons. First, because most mutations are recessive, they are not expressed if they are in the heterozygous state. Recessive mutations are hidden, or masked, if a good form of the gene is present in the same cell. Even if a parent's germ cell contains a faulty gene—one that specifies, for example, that a structural abnormality (misshapen limbs, perhaps) should develop—if it is recessive, it is likely to be overridden in the offspring by the normal gene from the other parent's germ cell. Only if both genes contain the same mutation will abnormal growth occur (as occurs when the mutations are homozygous). As we have mentioned, because recessive mutations are hidden in diploid cells, they tend to accumulate. The second reason heterozygosity and homozygosity are important is because of heterozygote superiority. For some genes, the heterozygote is just plain superior to either of the homozygotes. We will discuss heterozygote superiority further in just a moment.

Outcrossing Masks Mutations

To understand the purpose of outcrossing in diploid organisms, we need to consider how these organisms can repair their genes without

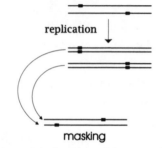

Figure 2. Coping with mutations

creating too much homozygosity when they replicate their DNA and reproduce. In Figure 2, deleterious mutations for two different genes (solid bars) are shown masked in the heterozygous state. For simplicity, only one chromosome pair is shown (the actual number of chromosomes varies widely—a human has 23; a toad, 11; a horse, 32; a dog, 39; a mosquito, 3). To reproduce, the cell must first replicate its DNA, which process results in a pair of identical chromosomes called "sister chromosomes." Because mutations can be replicated, sister chromosomes contain exactly the same mutations. To divide, the cell must put two chromosomes in each of the two daughter cells. It is easy to see how this still maintains heterozygosity for the mutations: pick one chromosome from each pair of sister chromosomes. This is basically what happens during mitotic cell division. Thus, it is easy for a diploid cell (or organism) to reproduce without expressing its hidden mutations as long as it doesn't need to make recombinations. In a similar way, homozygosity can be avoided for loci at which the heterozygote is superior.

What happens if there is damage in need of recombinational repair, in addition to hidden mutations? After the cell replicates its DNA, there will be a gap across from the damage (recall that damage can't be replicated), while the mutations will be replicated as they were in Figure 2. The cell must now repair the damage by recombination using the second chromosome. As we have learned in the last chapter, recombination will sometimes result in crossing-over (depending upon how the Holliday junctions are resolved, as shown in the double-strand break repair model). As a result of crossing-over, mutations may change position from one chromosome to another. In

Figure 3, one of the mutations has changed position. Whether this occurs depends on chance; the cell cannot determine beforehand which, if any, mutations will change position. Since its enzymes cannot find the mutations (after all they can't even recognize them), there is no way the cell can know where the mutations are located, that is, which chromosome they are on, after a recombinational repair event. Consequently, there is no way the cell can pick two chromosomes to give to its offspring and be sure they contain heterozygous mutations. (In the example shown, there are six possible ways to pick the two chromosomes, and two of them, or one-third, produce homozygosity for one of the mutations. However, the exact probability of unmasking mutations by making them homozygous depends

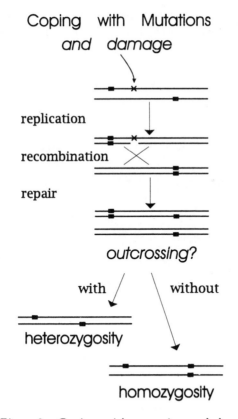

Figure 3. Coping with mutation and damage

on how many mutations there are, as well as on other factors.) My main point is that hidden mutations are expressed if recombinational repair occurs without outcrossing. In a similar way, other genes, such as those that have superior heterozygotes, will become homozygous if recombination occurs without outcrossing.

Although outcrossing doesn't affect the nature of the chromosomes produced by meiosis, it drastically improves the chances of keeping genes heterozygous (as long as the parents aren't genetic relatives). Germ cells from different parents usually contain mutations in different genes, so the chances of homozygosity are small. Although all organisms carry mutations, unrelated parents carry different ones. One parent may carry a recessive mutation that, if homozygous, would disrupt the functioning of its heart, and the other may carry a recessive mutation that would disrupt the functioning of its brain. When these parents mate, their mutations continue to be masked in their offspring. Because of the need to both repair damage and avoid homozygosity, recombination with outcrossing, sex, is often the best strategy.

Mating with Oneself

Relatives share genes, and, in particular, they share the same deleterious mutations. Incest, or mating between relatives, is bad for offspring because the offspring are likely to be homozygous: for example, they might receive the same deleterious mutation from each of their parents. Consider a brother and sister. Each may have inherited the same recessive mutation from their dad, say a mutation that disrupts normal development of the brain (there are known to be many such mutations). They are not affected, because they each received a normal gene from Mom. But if the brother and sister were to mate, the combination of their germ cells would have a one in four chance of containing the same brain-disrupting mutation received from Dad. There are several such lethal recessive mutations, and many more that are less severe. Offspring produced by such an incestuous mating would have a good chance of expressing some lethal or deleterious gene and not surviving.

If a female reproduced alone, using meiosis for gene repair but not outcrossing with a male, it would be as if she mated with herself. And if mating between relatives is bad for offspring, imagine the consequences if a mother mated with herself. Not a good idea, is it? Nor should it ever occur. But it does in many parts of the animal and plant kingdoms. Evolution shouldn't have invented selfing, but it has, many times. We must have left something out of the picture. But, what? It turns out that there are, in fact, benefits to selfing and mating with relatives, but before we consider them, let us get a little background on selfing and other forms of incest so that we may better understand the purpose of outcrossing.

Selfing, or mating with one's self, is an extreme form of inbreeding. Other common forms of inbreeding include brother-sister mating and parent-offspring mating. Such incestuous matings occur regularly in some species of fungi, plants, insects, spiders, fish and even mammals. Approximately 17% of all plant species have selfing as their predominant mode of reproduction. We will focus on selfing as the most extreme example of inbreeding, but keep in mind that what we learn about the advantages and disadvantages of selfing applies to all organisms that inbreed regularly.

Is selfing sex? Selfing can be thought of as a kind of asexual reproduction since there is no outcrossing. However, it differs from simple cloning, as occurs in bacteria or during the vegetative propagation of a strawberry plant, because of meiosis. When organisms self, they undergo a normal meiosis, with recombination and crossing over, to produce germlike cells. The diploid state is reinstated by fusion among these germlike cells (taken from the same organism) instead of among germ cells from different individuals as occurs during outcrossing.

Outcrossing is not an all-or-none phenomenon. A brother and sister are different organisms, but genetically speaking they are one-half the same, because they share 50% of their genes. Incest between a brother and sister is, of course, sex, even though it involves parents that are half the same. Selfing can be thought of as mating between parents that are completely the same, genetically speaking. Is selfing really any different from brother-sister mating, except for the matter of the degree of genes shared between the parents? No, selfing is an extreme form of inbred sex. To understand why outcrossing sex is maintained in most sexual populations we need to understand more about selfing and inbred sex.

Benefits of Selfing

In our previous discussion of the advantages of outcrossing, we depended heavily on the fact that recessive mutations accumulate in diploid populations. However, deleterious mutations accumulate only if they are not expressed, and for them to not be expressed, it helps for there to be outcrossing. Outcrossing promotes the accumulation of recessive mutations, which in turn promotes more outcrossing. In short, outcrossing promotes outcrossing. Is there no way out of this circle?

If only all the mutations could be purged from the population, then recombinational repair without outcrossing might be possible and the costs of mating could be avoided. Indeed, this must happen in the evolution of selfing and other forms of inbred sex. These organisms have found a way of abandoning the outcrossing aspects of sex but keeping meiosis with its effective means of gene repair. Because inbred forms of sex, like selfing, are fairly common, we know that the purpose of outcrossing is not an absolute need and may be avoided in certain situations.

What exactly are the benefits of selfing? Both selfing and outcrossing parents enjoy the advantage of gene repair during meiosis. However, a female that mates with herself has one big advantage over one that outcrosses and mates with a male—she doesn't require a mate, and, furthermore, she contributes all her offspring's genes (although in more homozygous combinations). These advantages of selfing are the flip side of the costs of sex mentioned in the first chapter, and they have not yet been included in our discussion of selfing.

In Chapter 1 we noted that the genetic costs of sex stem from two parents having to share genes in making their offspring—each parent contributes one-half the offspring's genes, in contrast to a selfing parent that contributes all the offspring's genes. This means that a selfing parent starts out twice as fit at passing on genes as does an outcrossing parent. This is the two-fold benefit of selfing that we seek.

Inbreeding Depression Rule

As we know, there is also a cost to selfing, because the offspring tend to be homozygous. The depression in fitness that results from selfing and other forms of inbreeding has been well documented and is termed "inbreeding depression." We might expect that if the cost of selfing, as measured by inbreeding depression, is less than one-half (the cost of outcrossing, which is the converse of the twofold benefit of selfing), selfing should win over outcrossing. Conversely, we predict outcrossing to win when inbreeding depression is greater than one-half. Unfortunately, matters are not so simple.

The amount of inbreeding depression created by selfing depends on many factors, including the mutation rate and the previous history of the population (whether it has been inbreeding for long or has recently switched). Inbreeding depression is also caused by heterozygote superiority. Inbreeding depression is discussed in greater detail in the Chapter Notes.

As we have mentioned, for some genes heterozygotes are just plain superior to homozygotes, for reasons other than their ability to mask deleterious recessive mutations. For example, two different alleles may produce two different enzymes that when combined together work better than either enzyme can work alone. Heterozygote superiority can make a substantial contribution to inbreeding depression. When heterozygotes do better just because they mask mutations, the advantage of the heterozygote (fitness of *Aa*) is no greater than the advantage of the homozygote that has been purged of mutations (fitness of *AA*). However, if the heterozygote is intrinsically superior, the heterozygote has greater fitness than both homozygotes (AA and *aa*)—any homozygosity reduces fitness (and the purging of mutations discussed in the Chapter Notes can no longer help selfers). Mathematical work has shown that only a few gene loci need exhibit heterozygote superiority for there to be a substantial advantage to outcrossing.

A particularly well studied example of a gene with heterozygote superiority is the mating type locus in the common yeast *Sacchar-moyces cerevisiae*. An example of a gene that exhibits heterozygote superiority in humans is the sickle-cell gene common in black popu-lations in West and Central Africa, where it reaches frequencies of 16% (it is present in about 5% of black Americans). The gene affects the hemoglobin molecule which serves to transport oxygen in blood and causes red blood cells to take on a sicklelike shape. Children homozygous for this gene usually die in infancy. When the gene is present in heterozygous state, the capacity of the hemoglobin mole-cule to transport oxygen is not affected, but other characteristics of the red blood cell are changed for the better. It had been noticed for some time that the sickle-cell gene reaches highest frequencies in areas of Africa in which malaria is common. As it turns out, the malaria parasite has difficulty infecting red blood cells of sickle-cell heterozygotes. This protection from malaria when the gene is in the heterozygous state is something neither homozygote can provide, and this advantage keeps the sickle-cell gene in high frequency.

The inbreeding depression rule seems simple enough, but, unfor-tunately, its predictions don't always work: at least that is what recent computer and mathematical models have told us. To under-stand why, we need to consider how natural selection operates on sex genes in more detail, and this is done in the Chapter Notes. We discover there that complex associations develop between sex genes (genes that change the mating system) and nonsex genes. These genetic associations tend to invalidate the inbreeding depression rule. The rule is a poor guide to how natural selection operates on out-crossing and selfing. For example, selfing may increase even though inbreeding depression is greater than one-half, if it is caused by recessive mutations. If inbreeding depression is caused by heterozy-gote superiority, outcrossing is favored over selfing even though in-breeding depression is less than one-half.

Even if the population has been inbreeding for some time and, by the death of especially mutated organisms, has purged many of the recessive mutations, new ones are occurring all the time. As already mentioned, in the genome of the fruit fly, approximately one new mutation occurs every generation. More complicated organisms probably have more new mutations. There is a benefit to masking these new mutations, just as there is to masking old ones that have accumulated. Even if a species has found a way to evolve selfing as a means of repairing DNA, mathematical and computer work shows that the benefits of heterozygosity can, under certain conditions, lead the species back to outcrossing.

Diploid organisms reproduce in diverse ways. Selfing is just one of several modes of reproduction that accomplishes meiotic gene repair in a "closed system," that is, without outcrossing. Other examples of such closed reproductive systems go by the names "automixis" (found in mites, spiders and insects), and "endomitosis" (found in lizards). As we have learned, selfers do not pay the costs of sex but run the risks associated with homozygosity. We have mentioned how selfing is an extreme form of inbred sex, similar in kind to other inbred mating systems that commonly occur throughout the living world, such as brother-sister mating and parent-offspring mating. There are also modes of reproduction, like cloning and vegetative reproduction, that are effective at masking recessive mutations but unable to effectively repair damaged genes (recall Figure 3 on p. 123). How much of this variation in mode of reproduction can be understood in the context of genetic error, the need to repair genes, and cope with deleterious mutation and the harmful effects of homozygosity? This is the question we turn to now.

Mutation Load at Equilibrium

Although genes now come packaged in cells inside organisms, it is sometimes helpful when discussing genes to think of them as existing alone. We can imagine all the genes in a population of organisms as being removed from the organisms and existing by themselves in a kind of gene pool, much as we imagined they once were early in the history of life before cells were invented.

Deleterious mutations comprise the sludge of the gene pool. Although individually random and inherently unpredictable, mutations must occur. As the French biochemist Jacques Monod phrased it, "Drawn out of the realm of pure chance, the accident enters into that of necessity." And so the sludge accumulates relentlessly with time. As each new mutation is added, the well-being of the population, and of the organisms that comprise it, goes downhill. The loss in fitness that results from the sludge is termed the mutation load. With time, the load of mutations would become so unbearable that the population would go extinct, were there not some way to keep their numbers in check.

The rate of mutation can be reduced by replicating DNA carefully and avoiding chemicals and agents that harm DNA, but the rate cannot be reduced to zero. Some mutation is unavoidable and, although recessive mutations accumulate slowly, with time their numbers build up. Eventually they must be reckoned with.

Lower survival, having fewer offspring and even death are some prices organisms pay for carrying deleterious mutations. Ultimately, as mutations are expressed, either through an effect in heterozygous state or because of homozygosity, natural selection punishes their carriers by removing their genes (and their mutations) from the gene pool. This is good for the population's gene pool (it is cleansed of the deleterious genes) but bad luck for the organisms involved. As new mutations are created by DNA replication, they must be removed by natural selection. Mutation in, mutation out. At some point, at equilibrium, a steady state is reached at which the input of new mutations equals their removal from the population.

The great English biologist J.B.S. Haldane argued in 1937 that, at equilibrium, the mutation load depends primarily on the rate at which new mutations occur (on a per haploid genome basis), μ, not upon how deleterious the mutations are. Slightly deleterious mutations accumulate to greater extent than the nasty ones, so the overall effect on the gene pool, as measured by the mutation load, is the same.

Consider just the lethal mutations in a gene pool of size $2N$ (N diploid organisms gives $2N$ genes in the gene pool). If μ is the mutation rate per haploid genome, $2N\mu$ new mutations are added every generation in a diploid population. At equilibrium, the number of mutations eliminated by natural selection equals the number of new mutations added by replication errors. We can think of the mutation load, L, as the fraction of the population that must die to eliminate these mutations. If mutations are recessive, every death will remove two mutations. At equilibrium, the number of mutations eliminated ($2NL$) must equal the new ones added ($2N\mu$), so the mutation load equals the mutation rate ($L = \mu$). This is Haldane's principle.

What effect does the mode of reproduction have on mutation load? It turns out very little, at least if the population size is large and if mutations don't interact with each other in determining fitness. In these situations, the mutation loads are about equal for sexual and asexual populations. Indeed, all modes of reproduction have approximately the same mutation load at equilibrium, when the input of mutations is balanced by their removal from the population. As Haldane first proposed, the mutation load depends primarily on the mutation rate. This is no longer true if the populations are small or if mutations interact drastically in their effects on fitness as discussed in Chapter 5; in these cases, the mode of reproduction matters, and in certain situations, sexual populations do better.

Numbers of Mutations Maintained at Equilibrium

Although different modes of reproduction have about the same average fitness at equilibrium, they harbor vastly different numbers of recessive mutations. Vegetative, or clonal, reproduction harbors large numbers of recessive mutations but expresses them rarely. In contrast, inbreeding forms of sexual reproduction, such as selfing, harbor relatively few mutations but express them quite often. The net result is the same: the mutation load depends only on the mutation rate μ and not on the mode of reproduction or on how deleterious the mutations are.

Results of mathematical and computer studies of different modes of reproduction in diploid organisms are shown in Table 1. The masking ability is given in the second column in terms of the number of recessive mutations maintained in the population at equilibrium. The greater the masking ability, the more mutations maintained, hidden in the chromosomes waiting to be expressed. There are substantial differences between modes of reproduction.

Table 1 Modes of diploid reproduction

Reproductive system	Masking Ability (number of lethals)	Recombinational Repair	Source of Backup Chromosome
automixis	low ($\approx 2\mu$)	yes	self
selfing	low ($\approx 2\mu$)	yes	self
outcrossing	intermediate (≈ 10 to $\sqrt{\mu n}$)	yes	another individual
endomitosis	high ($\approx n$)	limited	self
vegetative	high ($\approx n$)	no	not applicable
apomixis	high ($\approx n$)	no	not applicable

The parameter n is the number of genes, which in higher organisms is 50,000–100,000.

Those that are effective at masking recessives accumulate many mutations; those that are ineffective accumulate few.

While the effect on fitness of mutational load is the same, the different numbers of mutations leads to an advantage of changing the mode of reproduction and moving down the table. However, the need for recombinational repair of damage (given in the third column) favors modes of reproduction higher up in the table. The outcome of these conflicting needs—recombinational repair and masking mutations—is outcrossing sex. Let us consider an example.

Evolution of Outcrossing

The rather limited data from the common fruit fly *Drosophila* suggest that the mutation rate per haploid genome, μ, equals about one-half for lethal and nonlethal mutations combined (recall Figure 1). This yields an equilibrium fitness of about $\exp(-\mu) = 0.6$ (Previously, we used $1 - \mu$ for the average fitness of the population, but for technical reasons $\exp(-\mu)$ is more accurate.) The mutational load is not so large as to be overwhelming. If it were much smaller, we might expect that selection would favor an increase in the number of gene functions, n; insofar as μ should be proportional to n, selection for increased n would increase μ. Since the effect of the mutation rate, μ, on fitness is the same for all reproductive systems, we do not expect different reproductive systems to have radically different mutation rates. Consequently, we assume that the mutation rate is unity for all diploid reproductive systems ($\mu = 0.5$) and consider the effect of changing the mode of reproduction on the expression of recessive mutant alleles.

Let us reconsider the evolution of outcrossing sex and selfing from this point of view and ask if natural selection might be expected to change one mode of reproduction into the other. First consider a selfing population in which all the organisms mate with themselves. At equilibrium the population of selfers has fitness $\exp(-\mu)$. Few recessive mutations will have accumulated, but new ones occur each generation. A new kind of organism that outcrosses by mating with another organism will mask *all* of these recessive mutations in its offspring because the two mates will have different recessive mutations. This new sexual kind of organism will have a fitness of unity. The sexual outcrosser must, however, pay the costs of sex, C, so that the overall fitness of outcrossers is $1/C(C>1)$. Consequently, if $C < \exp(\mu)$, then a sex gene for outcrossing should increase. Using the value of $\mu = 0.5$, $\exp(\mu) = 1.7$, and so the costs of sex must be less than about 1.7. This condition allows for fairly large costs of sex—approaching the twofold value used in the inbreeding depression rule above. So let us assume that this condition is satisfied (sometimes the costs of sex are less than the twofold value) and that outcrossing starts increasing in the population and that selfing becomes less common.

After outcrossing becomes the common mode of reproduction, the number of recessive mutant alleles will continue to increase beyond what they were under selfing. They will accumulate until the outcrosser has an equilibrium fitness of $\exp(-\mu)$ due to mutation load, the same as the selfer had initially. However, the outcrosser's total fitness is actually less that what the selfer had, since the outcrosser's fitness is reduced both by the costs of outcrossing and by its mutational load ($\approx\exp(-\mu)/C$). Consequently, the overall effect of the transition from selfing to outcrossing has actually been a reduction of fitness and the well-being of the population! Outcrossing sex must evolve, but when it does the population is less well off than when it was selfing. Sex is a necessary, but less than optimal, solution to the conflicting needs of repairing DNA damage and masking deleterious mutations.

Although the population is less well off, once outcrossing sex has evolved, the reverse transition back to selfing, or other inbreeding forms of sex, is inhibited by the unmasking of the many new recessive mutations that have accumulated during the passage to outcrossing. Hence, among the first four cases in the table, which are those with complete recombinational repair, outcrossing is favored by natural selection.

Outcrossing Keeps Errors Uncorrelated

Let's sum up. We have found that the need to cope with both kinds of errors, damage and mutations, can lead to the evolution of sexual reproduction in a diploid population. Gene repair continues to be the purpose of recombination, as it was in simple haploid viruses and bacteria, but the purpose of outcrossing has shifted from repairing damage to coping with mutations and other deleterious effects of homozygosity. In simple haploids, outcrossing is required to provide a second DNA molecule for gene repair. In diploids, however outcrossing no longer serves this function. Instead, the need to cope with deleterious recessive or nearly recessive mutations helps maintain outcrossing. Viewed more generally, outcrossing has a similar function in both haploids and diploids. In both cases, outcrossing functions to provide the organism with good DNA at the site of errors; at the site of damages (in the case of haploids), and at the site of mutations (in the case of diploids).

We now have the components of a complete theory of sex which are completely general: the need for gene repair (discussed in Chapter 6) together with the need to cope with the deleterious effects of homozygosity—the expression of recessive or nearly recessive mutations and the lower fitness of homozygotes at loci for which the heterozygote is superior. In the next chapter, we put these ideas together and consider the complete theory, its strengths and its limitations.

◆ 8

Plato's Theory

Science has so little to tell us about the origin of sexuality that we can liken the problem to a darkness into which not so much as a ray of a hypothesis has penetrated. In quite a different region, it is true, we do meet with such a hypothesis; but it is of so fantastic a kind—a myth rather than a scientific explanation—that I should not venture to produce it here, were it not that it fulfills precisely the one condition whose fulfillment we desire. For it traces the origin of a {sexual} instinct to a need to restore an earlier state of things. . . . *What I have in mind is, of course, the theory which Plato put into the mouth of Aristophanes in the* Symposium, *and which deals not only with the origin of the sexual instinct but also with the most important of its variations in relation to its object. . . . Shall we follow the hint given us by the poet-philosopher, and venture upon the hypothesis that living substance at the time of its coming to life was torn apart into small particles, which have ever since endeavored to reunite through the sexual instincts? That these instincts, in which the chemical affinity of inanimate matter persisted, gradually succeeded, as they developed through the kingdom of the protista, in overcoming the difficulties put in the way of that endeavor by an environment charged with dangerous stimuli—stimuli which compelled them to form a protective cortical layer? That these splintered fragments of living substance in this way attained a multicellular condition and finally transferred the instinct for reuniting, in the most highly concentrated form, to germ-cells?— But here, I think, the moment has come for breaking off.*

—Sigmund Freud
Beyond the Pleasure Principle

To be competitive, organisms must rapidly replicate their DNA while at the same time avoiding mutations and repairing damage. Passing on healthy genes is an essential component of fitness. Organisms more able to repair genes and cope with mutation will enjoy an advantage in the struggle to survive and reproduce. Maintaining the well-being of genes is basic to life.

Good Backups for Repair

Mutations and damage pose distinct problems that are solved by sex. It is crucial for gene damage to be recognized and repaired, and this can happen only if a good backup copy is available in the cell to replace the lost information. Repair of damage often requires recombination between two chromosomes in a diploid cell. Gene repair alone does not require sex. But gene repair requires that the backup copy be good information, and this is where sex fits in.

Many mutations are recessive, or nearly recessive, and tend to accumulate in diploid cells. Recessive mutations lurk beneath the surface of a perfectly normal phenotype, hidden behind the good gene copy at the same locus—out of sight but not out of the way. For what happens if the good copy becomes damaged and the backup is needed for repair? The repair enzymes won't know any better if the backup is mutated (as we know, mutations can't even be recognized). The enzyme mechanics will dutifully recombine the defective information at the damaged site. One kind of error will be converted to another; each is equally disastrous for the organism.

Have you ever had a flat tire and opened the trunk of your car only to discover that the spare was also flat; or reached for the backup copy of a damaged computer file only to find that it, too, was bad? Managing error and its propagation through informational networks, whether computer based or genetic, is serious business and critical for maintaining the functioning of these systems.

It is ironic that the very redundancy needed for gene repair provides an opportunity for its own undoing, as recessive mutations accumulate. Some expression of recessive mutations is helpful to keep their numbers in check, thereby avoiding a diploid cell that has become functionally haploid by the accumulation of a recessive mutation at each and every locus. Both recombinational repair and masking of recessive mutations require diploidy, but diploidy alone is not a complete strategy; diploidy must be combined with outcrossing, as occurs during sex.

Recalling Gene Damage and Mutation

Let's review what we have learned about damage and mutations from earlier chapters. Genetic damage is a physical alteration in the structural regularity of DNA. The DNA molecule may become broken, or chemical cross-links may physically bind the two DNA strands together and prevent them from replicating or expressing their information. Nucleotides, the characters of the DNA alphabet, may also become damaged; their defining bases, referred to as *A, T, G* or *C*, may become chemically modified so that they no longer can be copied. Or, they may simply fall off the nucleotide backbone, leaving no characters at all. Such damage interferes with DNA replication or transcription of DNA into RNA—a critical step in making proteins. Damage is neither replicated nor inherited, but it can be recognized and repaired in the DNA molecule if a good backup copy of the gene is available in the cell.

Excision repair enzymes remove single-strand damage, after which the undamaged strand serves as a template for replacing the single-strand gap; recombination need not occur. Nevertheless, recombination is effective at repairing single-strand damage. Damage that affects both strands at the same or nearby positions, double-strand damage, cannot be repaired in this way since both strands of the DNA duplex are damaged. This double-strand damage can only be repaired by recombination and only if genetic redundancy exists, as occurs in diploid cells. As we have learned in Chapter 6, studies in bacteria and yeast show that double-strand breaks and cross-links can be efficiently repaired by molecular recombination between two homologous chromosomes. Substantial evidence supports the view that double-strand damage is prevalent in nature, has a significant effect on fitness and can only be repaired by recombination between homologous chromosomes.

Damage can be repaired while mutations cannot. Mutations are changes in the base pair sequence of DNA resulting from substitution, addition, deletion or rearrangement of the standard base pairs. Mutations do not generally alter the physical regularity of the DNA molecule, as it remains a regular sequence of A, T, G and C's. Consequently, mutations are replicated and thus inherited, but they cannot be recognized in the DNA molecule. Because mutations cannot be recognized in the DNA, they cannot be repaired, and they tend to accumulate in cells. Mutations are "recognized" only after they are expressed in the phenotype and their effect on fitness is evaluated by natural selection. Sex can help mask mutations and it can also help make natural selection more effective at getting rid of mutations.

Coping with DNA Damage

For higher organisms, recombinational repair occurs primarily during meiosis, although a low level of recombinational repair occurs during mitosis. Certain means of reproduction, especially in plants, do not have meiosis—like vegetative reproduction (as occurs, for example, in the propagation of a strawberry or grass plant by runners) or apomixis (a way of producing seed by asexual means in, for example, a dandelion plant). Not having meiosis, these organisms must cope with gene damage by some means other than recombinational repair. This leaves cellular selection as the primary way these organisms must deal with double-strand damage.

Cellular selection occurs when damaged cells die and are replaced by replication of undamaged cells. In this way, a population of cells may be able to persist and cope with damage (for example, a population of cells in a root tip of a plant, or in a tissue like the epithelial tissue lining the stomach). For this strategy to work, some cells must remain undamaged and resources must be available for them to replicate. In vegetative reproduction of a plant by runners, many somatic cells are committed to the production of each new plant; in apomixis, a single cell is committed to each offspring. Consequently, vegetative reproduction should not be as vulnerable to gene damage as is apomixis.

Recombinational repair is the strategy of choice when the occurrence of damage is so high that most cells contain damaged genes or if resources are not available for replication of undamaged cells. There are two potential sources for the backup chromosome needed in recombinational repair. In a closed-system strategy, the chromosome comes from the same parent or close genetic kin. In an open-system strategy, as occurs during outcrossing, it comes from another, unrelated individual.

If the only problem were gene damage, the most effective strategy would be a closed system such as selfing or automixis (virgin birth involving an ordinary meiosis that is followed by an internal process for restoring diploidy), since these modes of reproduction would avoid the major costs of sex that stem from outcrossing. However, the deleterious effects of homozygosity can lead the species to an outcrossing mode of reproduction. Deleterious recessive mutations occur continuously and are expressed by recombinational repair. Loci at which the heterozygotes are superior also make significant contributions to inbreeding depression and aid the movement during evolution towards outcrossing sex from closed systems of gene repair.

Coping with Deleterious Mutations

Mistakes made in the replication of DNA produce mutations. Since improvements in the accuracy of DNA replication has costs, there are cost-effectiveness barriers to the indefinite improvement in the fidelity of DNA replication. A nonzero spontaneous mutation rate is therefore unavoidable. Deleterious mutations have a much greater effect on fitness in the short run than do favorable mutations because of their much higher rate of occurrence.

Little can be done concerning the input of new mutations: however, their removal from the population can be enhanced by sex. First, offspring produced sexually have different numbers of deleterious mutations, while asexually produced offspring have the same number their parents had, plus any new mutations that might have occurred when their parents reproduced. The greater variance in mutation load of sexually produced offspring can be important in both small and large populations. The ratchet discussed in Chapter 5 occurs primarily in small populations. By avoiding the ratchet, a sexual population can keep a check on its numbers of mutations. The ratchet undoubtedly sets broad limits on the size of the genome that asexual organisms could attain. If mutations interact synergistically, that is, if the deleterious effect of each additional mutation is greater and greater, the average fitness of sexual populations may be greater than asexual populations because more mutations are eliminated each time an organism dies. An extreme form of synergism, termed threshold selection, was discussed in Chapter 5 to illustrate how a sexual population can do better than an asexual one.

Although sex helps mask mutations, by chance, two parents may carry in heterozygous state deleterious recessive mutations for the same gene. The resulting offspring could be homozygous and would then suffer in fitness. Unmasking of deleterious mutations happens rarely among outcrossing parents, especially when compared to its likelihood among selfing parents, but it does occur, as we know from the expression of genetic diseases in human populations. Although resulting in much suffering in our own species, this unmasking helps reduce the number of recessive mutations to a theoretical estimate of about ten (see Table 1 on p. 132). This is far less than the theoretical maximum number of recessive mutations that can accumulate in asexual mitotically reproducing organisms, which is around 100,000 mutations in every cell. (From inbreeding studies, the number of recessive lethal mutations in humans is thought to be around two or three, less than the number ten calculated in the previous chapter. Other factors, such as inbreeding, mate choice, or incomplete recessivity must be involved in reducing their number below the theoretical estimate of ten.)

Why Sex Evolved: An Overview

In summary, there are two basic errors in the transfer of genetic information: gene damage and mutation. Recombination universally serves to repair gene damage. The two principal features of sex, recombination and outcrossing, both originated in bacteria-like creatures (those made up of just a single haploid cell) for the purpose of repairing gene damage. Recombination and outcrossing are maintained in more complex, many-celled creatures by the advantage of repairing damage while not expressing deleterious mutations. The deleterious effects of homozygosity at loci for which the heterozygote is superior also helps maintain the outcrossing aspects of sex in these complex organisms. Although the need for gene repair does not by itself explain the need for outcrossing sex in complex organisms, it does, by way of its solution in terms of recombinational repair in a diploid cell, set the stage for the accumulation and expression of deleterious mutations and the generation of homozygosity for genes that have superior heterozygotes. It is the problem of accomplishing gene repair while coping with the deleterious effects of homozygosity that maintains outcrossing sex in diploid organisms.

According to the gene repair mutation view, sex is a *cooperative* venture between mates. Evolutionary discussions of sexual behavior often emphasize the antagonistic and aggressive interactions between the sexes. Sexual behavior among males and females is indeed risky, as we have illustrated many times in previous chapters. However, these antagonistic interactions stem from the different reproductive strategies pursued by males and females. They do not stem from sex itself, because sex did not originate for the purpose of reproduction. The association of sex with reproduction came much later in the history of life. As we found in Chapter 3, the male and female sexes evolved when anisogamy, different-sized gametes, was invented. For over a billion years, perhaps even 2 billion years, of life's history, while sex involved the fusion of same-sized and equal gametes, the mates existed in relative harmony with single-minded purpose, that of coping with gene error.

As with all cooperative behaviors, it is possible to imagine cheaters in the gene repair sex game. My colleagues and I have studied the possibility of several different kinds. We imagined that one kind of cheater might practice "selfish sex," in which it has sex only if it itself is damaged and never when it is healthy. Another kind of cheater might practice "parasitic sex," in which damaged cells would "prey" on healthy cooperative cells. Happily, there are good reasons why these selfish parasites don't evolve and why we expect sex to maintain its cooperative nature—at least until males and females were invented.

When George Williams wrote the book that reopened the problem of sex, he noted that if sex wasn't so common, evolutionary biologists wouldn't wonder why it hadn't been invented. This is another way of admitting that the imagined benefits of sex, at least those associated with hopeful recombinations, are not very persuasive and powerful. We wouldn't have seen the need for such long-term and iffy benefits unless we were forced to by the lack of sound theory for the ubiquity of sex.

The advantage of recombinational repair sets the problem of sex on far firmer ground. Could we imagine genes without a means of making repairs? Hardly—except under situations in which they can be thrown away. Could we imagine a way of making repairs without inserting spare parts? Hardly. Could we imagine a reserve of self-replicating spare parts that didn't accumulate errors of the mutational variety? Sometimes, but not very often. Recombination is necessary for life, and outcrossing is often selected for as a by-product of recombination. Sex was there in the beginning of life, it was reinvented when cells were invented, and it has been with life ever since.

Gene repair together with the deleterious effects of homozygosity explain features of sexual systems that other hypothesis don't even treat. However, these ideas also raise special problems. We now consider these related issues.

Sex in Many-Celled Haploids

All sexual creatures alternate between haploid and diploid cell stages. Some genes are concerned with the characteristics of both kinds of cells, but other genes are expressed in just diploid or haploid cells. Masking of recessive mutations occurs for genes that are expressed only during the diploid stage. Recessiveness of a gene is a moot point in haploid cells since there is just one copy of each gene. Similarly, heterozygote superiority cannot occur in a haploid cell.

Most of the 40,000 to 100,000 genes needed to make a complex organism are expressed only during the diploid stage. Complex organisms have specialized cell types that differentiate from one another by turning on different genes: liver cells, brain cells and others have different characteristics because they express different genes. Because the cells are diploid, recessive mutations are masked for these genes and they accumulate. However, for genes expressed during the haploid stage, recessive mutations do not accumulate since they are always expressed. For our own species this is a relatively small number of genes. For example, almost no genes are expressed in the human sperm cell; sperms are charged with energy from their parent cells in the testes. Likewise, the egg is provided with nutrients from diploid "nurse" cells.

However, in fungi and in some plants—for example, algae and stoneworts—the haploid stage predominates and the diploid stage is short-lived. Stoneworts are a special group of many-celled plants similar to algae. They are found in brackish water, rivers and lakes with hard water and are usually attached to the bottoms with root-like rhizoids. Best represented by the genus *Chara*, stoneworts are encrusted in lime deposits secreted by their cell walls, hence their name. Highly complex, the structures of the stonewort resemble higher plants rather than algae. In addition to their rootlike rhizoids, they have leaflike branches at regular intervals and upright cylindrical axes surrounded by a sheath of cells. All these structures are haploid. Sexual reproduction is achieved through the fusion of one large egg from the female sex organ (oogonium) and one small sperm from the male sex organ (antheridium).

There is a special problem in explaining outcrossing sex in many-celled haploid organisms like the plants just mentioned. For single-celled, as opposed to many-celled, haploids, outcrossing is needed to bring homologous chromosomes together into a cell for the purpose of recombinational repair. However, in many-celled haploid organisms, a template for repair is available in the next cell. Why not go there for the backup copy; why go to the trouble of fusing with a cell from another organism as occurs during sex?

A variation of this challenge concerns insects with the peculiar haplodiploid genetic system. In bees, ants and wasps, males are haploid and females diploid. Deleterious recessives are masked in females but are weeded out in males. It seems that outcrossing would no longer be needed to keep the number of deleterious mutations in check since most genes are expressed in both sexes and deleterious recessives would be weeded out in males. At first glance, the occurrence of sex in many-celled haploid and in haplodiploid species suggests that outcrossing may serve another function than avoiding homozygosity.

We must remember, however, that even in haploid organisms, there remain key gene functions that are expressed only during the diploid stage, such as the genes responsible for meiosis. For simpler organisms, such as fungi, meiosis is probably their most complicated activity since the many-celled haploid stage, although dominant in size, often has a modular construction with little differentiation. Furthermore, in mushrooms and bracket fungi, for example, the conspicuous structures are often dikaryons (heterokaryons in most other fungi) that contain two nuclei from different parents. These are functionally diploid (or polyploid), so they are capable of masking deleterious recessive alleles. This masking would give a selective advantage to outcrossing, as we have discussed in the previous chapter.

In stoneworts and haplodiploid insects, there are likely more genes concerned with aspects of the haploid stage since the haploid phases are complex structures containing several differentiated cell types. Because deleterious recessives would be expressed during the haploid stages, they should not accumulate. Inbreeding forms of sex, such as selfing, should then be a viable strategy for these organisms because there are fewer deleterious mutations to unmask. The data are broadly consistent with this expectation—haplodiploid insects and stoneworts are often inbred. As we expect, when there is less need to mask mutations or otherwise cope with homozygosity, there is less need for the outcrossing aspects of sex.

The Function of Premeiotic Replication

There is a peculiarity of meiosis that has gone virtually unnoticed by population biologists interested in the evolution of sex. Before meiosis begins, the DNA is replicated. Premeiotic replication of DNA is a general feature of meiosis in the vast majority of organisms. It leads to four homologous chromatids that then recombine in pairs.

Premeiotic replication has a definite purpose if we take the point of view of gene repair. Studies in the bacterium *E. coli* show that replication of DNA with single-strand damage leads to gaps in the new strands opposite the damage; furthermore these gaps promote recombinational repair. Such gaps have a molecular structure that is independent of the molecular structure of the original damage and can serve as a universal initiator of recombinational repair. Using gapped DNA as a signal for recombinational repair avoids the need of evolving a specific enzyme to recognize each specific kind of single-strand damage. This would be impractical anyway, except for the most common kinds because there are very many ways to damage a DNA molecule. Using a common signal for repair would be especially adaptive if naturally occurring damage were a mixture of many different kinds, with many kinds represented in low frequency.

Most Recombination is Cryptic

Recently, it has been shown that most recombinations do not result in new combinations of alleles. Most of the time, about two-thirds of the time in fruit flies and yeast, recombination during meiosis does not result in crossing-over. As discussed in Chapter 6, crossing-over refers to the exchange (from one chromosome to the other) of genes that reside to either side of the site of molecular recombination. Crossing-over must occur if the associations between genes are to change. In most recombinations, a short segment of DNA, less than one gene's worth, is patched from one chromosome to the other, leaving the vast majority of genes unaffected. Such patchwork recombination does not produce new combinations of genes, but it is an effective means of repair. Much of the time recombination is cryptic and no new gene combinations are produced. Cryptic recombination cannot generate new associations of genes at two or more loci. Thus the purpose of a majority of recombinations is not even addressed by the hypotheses presented in Chapter 5, although they are nicely explained by the need for gene repair.

Why Aren't All Recombinations Cryptic?

The fact that most recombinations are cryptic raises the question, Why aren't all recombinations cryptic? One could imagine an efficient repair process that didn't result in any crossing-over. If the recombination system could only be redesigned in such a way, this would reduce the need for outcrossing; little homozygosity for recessive mutations should be created by patching in a few hundred base pairs at the site of damage. One of my critical colleagues has called this strategy the "efficient-diploid-mitotic-repair-strategy." After all, we mentioned in the last chapter how recombinational repair can occur, though at low frequency, in mitotic cells. If hopeful variations weren't important for the evolution of sex and recombination, why not invent a more optimal form of gene repair than recombination plus outcrossing?

This is a serious challenge to the view of sex that I have presented and I do not have a simple answer. But there is an answer. In the long term, nothing is gained by switching to an efficient form of mitotic recombination. Furthermore, in the short term, the required evolutionary steps to get there are impractical. Let me explain.

There is an advantage to redesigning the recombinational repair system only if outcrossing is abandoned at the same time. There is no benefit to an outcrossing parent to redesigning recombination so that all recombinations are cryptic, because most mutations are already being masked. As we know from the last chapter, there is a large cost to giving up outcrossing, without redesigning the recombination system with no crossing-over, since doing so expresses the accumulated recessive mutations. For either redesigning recombination or abandoning outcrossing to be advantageous, they must occur simultaneously. In practice, it is quite difficult to evolve two specific changes in different complex structures at exactly the same time. In a selfing organism, there would be an initial benefit of avoiding crossing-over so that mutations would not become homozygous. However, the opportunity for this to occur is very short-lived. Mutations are expressed quickly and are either rapidly purged or the selfer goes extinct.

In practical terms, it would probably be very difficult to redesign the recombination system anyway. The molecular machinery of recombination is ancient. The basic enzymes first appeared in haploid, bacterialike creatures over 3,000 million years ago, and there has been a more-or-less continuous evolution of these proteins through bacteria and eukaryotes. Within bacteria, the nucleotide sequence similarity among recombination proteins is high, about 80% to 100%. The sequence similarity of recombination proteins between bacteria and phage and yeast is lower, about 30%, but the three-dimensional similarity of the shape of the proteins has been conserved. A leading research group on the molecular similarity of recombination proteins concluded recently that the "proteins in this group are members of a single family that diverged from a common ancestor that existed prior to the divergence of prokaryotes and eukaryotes."

The more ancient a structure or process is, the more entrenched it becomes in the organism, because other structures and processes that evolve later come to depend on it. For example, some biologists believe that the structure associated with crossing-over (called a chiasmata) helps the two chromosomes of a meitoic pair to separate and undergo chromosome disjunction. To give up crossing-over would then jeopardize the process of chromosome disjunction during meiosis. There are probably other features of meiosis that have come to depend on crossing-over, but these features are not the reason recombination evolved in the first place.

Even if the recombination system were redesigned with no crossing-over, in the long term nothing is gained as recessive mutations eventually build up and sabotage gene repair. Once the mutations accumulate, the recombination enzymes will patch in the defective information at the site of damage, creating a cell that is homozygous for the defective gene. The backup copy eventually deteriorates in any closed system such as "efficient-diploid-mitotic-repair."

The trick is to have efficient repair without letting recessive mutations accumulate till they fill up the genome. Outcrossing sex accomplishes this: so do inbreeding forms of sex, such as selfing. But as we have seen, inbred systems of repair can be unstable in the short term to the evolution of outcrossing. For these reasons, the efficient-diploid-mitotic-repair-strategy is no long-term strategy at all, and short-term steps that could lead a population toward it are impractical.

Variation Producers

Genetic variations are so basic to evolution that it is only natural to think that when a process produces variations this must be its purpose. The folly inherent in this way of thinking is clear when discussing mutation—DNA replication obviously doesn't exist to produce mutations! But it is easy to slip back into this way of thinking when discussing other kinds of variation producers such as plasmids, phage and recombination. Infectious elements such as phage, plasmids and transposons often reside in their host cell's chromosomes where they hide out until it is time to move. After they move, they insert themselves in the chromosomes of their new host. When they insert, they often create mutations and recombinations in their host's DNA. Some biologists have suggested that this may benefit the host by creating new variations. Yet it seems obvious in these cases that genetic variation is a secondary consequence of the primary need of the infectious element to move from host to host.

A similar argument holds for sexual recombinations. Recombination evolved for gene repair. To repair a machine a new part must be inserted for the damaged one, and the repaired machine may run differently as a result. Rarely, new adaptive combinations of genes are produced as a by-product of gene repair. These infrequent beneficial recombinants undoubtedly promote long-term evolutionary success of the species, just as infrequent beneficial mutations do. But they are not the consequence of sex that keeps individual organisms from becoming asexual.

All the diversity in the living world comes from mutations that are assorted and recombined into new forms by sex. However, sex did not evolve for this reason. The variation produced by recombinational repair may change the way an organism functions, but this is a secondary consequence of the need to keep the organism's genes healthy.

Risky Business

One expects random changes to a working machine to reduce its effectiveness rather than to improve it. The genome contains the instructions for building an organism which can be likened to an extremely complex machine. Random changes to these instructions, as occurs during mutation and recombination, likely produce organisms that don't work, rather than ones that work better.

Breaking DNA, as occurs during recombination, is risky business; the DNA molecule may not get put back together again. As testimony to the risk involved, many severe chromosomal aberrations stem from mistakes made during recombination. Should a parent take such risks in the hope of making its offspring work better—a parent, mind you, that is already working well? Parents have survived to adulthood and stand as evidence of a well-tested design. Why change the design in the off chance that an even better-designed offspring might be created? It doesn't make any sense, it really doesn't.

Breaking DNA is itself damage that must be repaired—hence the name "double-strand break repair model" for the mechanism of recombination that occurs during meiosis (presented in Chapter 6). Would organisms actively damage their DNA in hopes of producing something better? Certain plants are known to produce so-called secondary defense chemicals that protect the plants by damaging their predator's and parasite's DNA. According to the hopeful variation point of view, we should rethink the plants' strategy—perhaps this is a mutualism in disguise: the plants are actually trying to help their predators evolve faster!

To accept the risks inherent in breaking a DNA molecule, there must have been something wrong with the molecule to begin with—like damage that previously existed at the site of recombination. Damage increases the likelihood of recombination in all organisms studied. Even outcrossing has been shown to respond in an adaptive manner to DNA damage. Molecular biology is demanding a reconsideration of the function of sexual recombination. For example, the biochemist Michael Cox tells us that "if recombination (genetic diversity) is the primary mission of this system and represents the selective pressure for its evolution and maintenance, then many of the key properties of the RecA [recombination] system observed in vitro . . . become hard to rationalize."

The increase of recombination in response to DNA damage, the double-strand break repair model, the prevalence of cryptic recombination, the design of the recombination proteins, and premeiotic replication are all hard to reconcile with a view that sex evolved for the purpose of redesigning organisms. However, all these features of sex fall neatly into place once recombination is seen in its proper repair role.

We now understand why sex evolved. Let us turn our attention back to why it matters that sex evolved. The remaining two chapters take up two questions that Darwin dearly wanted to answer: Do organisms really matter in evolution? And why are there species? Again, we will find that it is sex that really matters.

◆ 9

Darwinian Dynamics

The expression often used by Mr. Herbert Spencer of the Survival of the Fittest, is more accurate, and is sometimes equally convenient.

—Charles Darwin
The Origin of Species

This survival of the fittest which I have here sought to express in mechanical terms is that which Mr. Darwin has called "natural selection, or the preservation of favored races in the struggle for life."

—Herbert Spencer
Principles of Biology

W ho are the fittest and why does it matter that we know? These are the questions I hope to answer in this chapter. I think you will be surprised by what we find out. We will discover that the answers to these questions depend critically, as does so much in biology, on sex and the way in which sex undermines the existence of organisms as meaningful objects in evolution.

Darwin's "Mistake"

I had just started my third year of graduate study in evolutionary biology, when, paging through an issue of *Harper's* magazine one

151

winter day in 1976, I noticed an article titled "Darwin's Mistake." According to Tom Bethell, a science writer, Darwin made a fatal mistake that undermined the entire theory of evolution by natural selection. "Who are the fittest?" Bethell asked. He found the answer in the standard texts of evolutionary biologists—the fittest are those that best survive. Upon substituting this definition of "fittest" into the phrase "survival of the fittest," we have the classic tautology "those organisms that survive best are those that best survive." Darwin's great principle is thus seen to be nothing more than an empty tautology, since there is no independent criterion of survival and reproductive success.

At first I laughed to myself. Could anyone really think that one of the world's greatest thinkers had written thousands of pages of careful analysis based on concrete biological examples but overlooked the fact that his underlying logic rested on nothing more than an empty tautology? Not likely, I thought. But where did the author go wrong? I was on the verge of devoting my career to studying evolutionary biology and wanted a simple answer. It should be easy to point out the flaw in Bethell's challenge, but where was it? As I began to think about "Darwin's Mistake," I found it raised some troubling questions—questions that I could find no clear answer to in my studies. Can we define fitness of an organism independent of its survival and reproductive success? Does it really matter?

Embarrassed by the tautology challenge and the lack of a simple response, biologists have argued desperately that, yes, they can measure an organism's fitness independent of the organism's survival and reproduction. Even contemporary philosophers of biology have joined them in support. As I studied these defenses over the years, I became more and more dissatisfied with them. The problem is with their focus on organisms. I have come to the conclusion that the emphasis on organisms implied by Darwin's claim is not only misleading but is entirely mistaken. We really can't measure an organism's fitness independent of its survival and reproduction, but we really don't need to. Because organisms are not what really matter in evolution.

Explaining Nature

The theory of evolution explains design in the living world. Until just over a hundred years ago this was the job of religion, not biology. Before then, for well over 5,000 years in Judeo-Christian and Moslem civilizations, it was God who explained life and adaptation—the fit of organisms to their environment. A watch implies a watchmaker, so argued William Paley in 1802 in *Natural Theology; or, Evidence of the Existence and Attributes of the Deity, Collected from the Appearance of Nature.* If the existence of a watch implies a watchmaker, well-designed organisms must have been designed by a Creator. The belief in a divine Creator of life has profound implications for the practicing biologist. In the seventeenth and eighteenth centuries, mainstream biology was primarily concerned with cataloging the links in the "great chain of being," that unbroken linear sequence of organisms in rank order—from the crudest infusorians up through plants, animals and eventually to humans, who, as humans viewed it, were just a notch below the angels who were right below God. Nature was seen as a highly structured system, a system created by God and one in which humans came out looking pretty good. Understanding how the theory of evolution can substitute for such unbridled wishful thinking is a daunting task. This task is now the job of philosophers of biology, as practicing biologists have shown less and less interest in epistemological matters.

The Easy, the Not-So-Easy, and the Surprising Answer

As we will see, there is an easy, a not-so-easy, and a surprising answer to the tautology problem, as this whole matter has come to be called by philosophers. The easy answer is that "survival of the fittest" is often false. In other words, less-"fit" organisms are often better at surviving and passing on their genes than are more "fit" organisms. Tautologies, being empirically empty, cannot be false; so clearly Darwin's phrase is not a tautology. While solving the tautology problem, the realization that Darwin's phrase can be false may do little to give us confidence in Darwin's theory, since we may wonder how Darwin could have laid such great emphasis on a falsity and still have a correct theory. Darwin was wrong, at least some-times, but it's okay. Understanding why Darwin was wrong, and why it's okay that he was wrong, reveals the true character of the process of natural selection that Darwin discovered.

The not-so-easy answer to the tautology problem is that fitness is just one component of natural selection. Survival of the fittest can be false because fitness is just one component of evolutionary success. Once we see how all the components of natural selection contribute to evolutionary success, we will see why less fit can be better. Al-though it may sound odd to you now, I hope you will agree after we are done that such an outcome is exactly what we should, at times, expect from Darwin's principles *when they operate in sexual populations*.

The surprising answer to the tautology problem is that organisms really don't possess fitness, at least not in the sense of an overall property that can be defined in a nontautological way. I say "surpris-ing" because during the past twenty or so years philosophers support-ive of evolutionary biology have been arguing just the opposite, that the concept of adaptedness, the overall fitness of an organism con-ceived of as a "propensity," provides an answer to the tautology problem. However, this notion of a propensity is a pure abstraction, of little relevance to how evolutionary biologists actually go about explaining nature. I hope to convince you that this purported "fit-ness" of organisms doesn't really exist, that is, fitness in the sense of an overall property or propensity of organisms that causes survival and reproductive success. Furthermore, it really doesn't matter that it doesn't exist, because individual organisms do not appear as ultimate products in evolution either. And all this because of sex.

"Survival of the Fittest"

But we have gotten ahead of ourselves. Let's look more closely at what Darwin and Spencer meant by "survival of the fittest." They wished to make a causal and predictive claim about the outcome of natural selection, in particular, that the property of "fitness" causes increased survival. But "fitness" and "survival" of what?

Organisms come to mind. Organisms with higher fitness survive better; that makes sense. But organisms just live and die. Organisms don't evolve by natural selection, only populations of organisms (or other kinds of replicators) evolve by natural selection. Darwin and Spencer must be referring to different kinds of organisms, defined by the traits they possess. For example, if we were thinking of the ancestors of present-day giraffes living on the African savanna, we might predict (noticing all the available food high up in trees) that a longer-necked giraffe would survive to reproduce better in the savanna than the kind with shorter necks. (Of course, what we really have in mind is the genes that confer upon the giraffe the trait of having a longer neck. These genes will increase, or so we predict.) As a result of their greater survival and reproduction, long-necked giraffes will increase in frequency and, with time, long-necked giraffes (and their underlying genes) will come to predominate in the population. This is evolution by natural selection.

Gene Survival

So returning to "survival of the fittest," we see that "survival" means not just survival of an organism through its life, since it must include how many offspring the organism has if it survives to reproduce, there being no point in surviving if no offspring are produced. Furthermore, to be predictive of natural selection, "survival" must include the likelihood that the genes of interest (the genes for long neck in the giraffe example) are passed on to those offspring. We are talking about sex here, and offspring receive genes from both parents. So the whole matter is quite complicated. But it should be clear that the bottom line in natural selection is survival of the genes that determine a kind of organism. The survival in question is not just from parents to offspring but survival over several, even many, generations. Gene survival is what evolution is all about. Of course, Darwin and Spencer did not know about genes, but this is how we must translate their phrase if we are to make sense of it in modern terms. A more descriptive phrase than "survival" of genes is "evolutionary success." Darwin and Spencer were saying that the "fittest" have more evolutionary success. Now what did they mean by "fitness?" That is more difficult.

One thing should be clear from our discussion of the tautology problem: fitness must mean something different than actual survival or evolutionary success. Whatever it is, fitness should be a cause of increased survival and evolutionary success—but it can't be the same thing. We need a criterion of fitness that is independent of actual survival and evolutionary success. That was the *Harper's* author's main point. Nevertheless, if things work out the way we expect— expect based on survival-of-the-fittest thinking, that is—then increased fitness should result in increased survival much of the time. So in practical terms, in those cases in which the fittest do survive the best, it may be useful to just measure fitness by its effect on survival (and reproduction). That is what is often done in practice. We can call this fitness notion "operational fitness" because it measures fitness by the operational criterion of actual survival and reproduction.

Operational Fitness

There are many kinds of operational fitness. The expected survival and reproduction of an organism, or a kind of organism specified by its genotype, is one kind (the so-called Darwinian fitness, W_{ij}, seen in many population genetics textbooks). For the most part, I have used "fitness" in this sense in the previous chapters. One simply counts the number of offspring a kind of organism produces and multiplies that number by the likelihood the organism survives to reproduce in the first place. But expected survival and reproduction of organisms doesn't include many aspects of natural selection, such as the mating system and sex. Another kind of operational fitness is the rate of increase of a genotype, usually expressed on a per capita basis, so that it becomes the rate of increase achieved by a single instance of the type. This is the notion of fitness used by the population geneticist Ronald Fisher, whom you have encountered several times so far in this book as one of the founders of modern evolutionary theory. Fisherian fitness is what is often measured by evolutionary biologists studying the common fruit fly, *Drosophila*, when they measure the rate of increase of a kind of fly in a cage in the laboratory.

It is important not to confuse Darwinian or Fisherian fitness (and other operational definitions of fitness such as "inclusive fitness") with their underlying causes. Let's consider a physical example for which matters are more clear cut, say, solubility in water. One way of measuring a sugar cube's solubility in water is, of course, to simply drop it into water and actually observe how long it takes to dissolve. This is like measuring fitness by seeing how well a particular organism actually survives and reproduces or how fast a *Drosophila* genotype actually increases in frequency in a laboratory cage. But no one would think that the *cause* of solubility is the same thing as its actual solubility. The cause of solubility lies in the object's molecular structure. Likewise, the cause of operational fitness must lie elsewhere.

Or consider the familiar example of whether a key fits a lock. Again, we could just try out the key to see if it opens the lock. That is the most direct approach. But we could also measure the shape of the key and measure the shape of the lock and figure out whether the key will open the lock without even trying the key. It is the physical shape of the key (and the lock) that determines whether the key will open the lock. We wouldn't think that the cause of whether the key opens the lock is whether the key opens the lock! Likewise, it would be wrong for us to think that operational fitness is anything more than a measure of something else. What is this something else? It must be a property or capacity of the organism's genotype, analogous to the molecular structure of the sugar cube or the shape of the key. But what?

Adaptedness

The word "adaptedness" is used by evolutionary biologists to refer to this elusive property. Biologists sometimes use the word "fitness" for both the concepts adaptedness (as an aggregate property of organisms or genotypes) and operational fitness (actual survival and reproduction or rate of increase). The basic distinction needed is between fitness as a rate of increase of a type over time (a type of gene or genotype) and fitness as the adaptedness of a type of organism. Clearly, adaptedness must be measured in a manner that is independent of actual survival and reproduction or the actual rate of increase.

What is adaptedness of organisms and how does evolution depend on it?

Simple mathematical models are helpful in answering this question as the models make explicit distinctions that are often implicit in verbal discussions about natural selection. A central point of our analysis will be that there is no adaptedness, at least not in the sense of a nontautological overall property of organisms that causes operational fitness. Instead, what we find in the mathematical statements of natural selection are "adaptive capacities," properties of the organism's genotype that in interaction with the environment cause operational fitness.

We have come far in our understanding of the effects of sex without having to make too much use of mathematical models. The questions and biological issues we have considered have been concrete enough that words have sufficed. We are now venturing into philosophy, where matters are more subtle and words often have multiple interpretations. For these reasons, I find philosophical analysis in need of quantitative treatment as scientific analysis in biology or physics. I see no other way to make my points than by using simple mathematical models. Please bear with me. First, an old friend.

Survival of the Fittest: False Because of Sex

If you have ever had a biology course at the college level, you likely considered the case of "heterozygote superiority" in the genetics or evolution part of the course. We have discussed heterozygote superiority several times already in our consideration of inbreeding depression and the evolution of outcrossing and selfing. Heterozygote superiority is one of the most worn examples in elementary population genetics. It concerns a single gene locus in a diploid organism. Organisms that are heterozygous (*Aa*) are assumed to be the most adapted and to have a greater likelihood of surviving to reproduce, than organisms that are homozygous (*AA* or *aa*).

The example of heterozygote superiority illustrates well why survival of the fittest, fittest in the sense of most adapted, can be false— if there is sex. In Chapter 7 (see also p. 205) we found that survival of the fittest could not explain the evolution of genes affecting the mating system if there was heterozygote superiority at a fitness-determining locus. Natural selection was complex in that example, as we were considering the interaction between sex and nonsex genes (mating system genes and fitness genes). We are now considering a far simpler situation of a single fitness-determining gene locus at which the heterozygote is most adapted and has the highest Darwinian fitness. In the notes to this chapter we consider this case in some detail. Is survival of the fittest true for the evolution of traits determined by these genes? The answer, again, is a resounding no.

The reason the fittest do not best survive is because of sex. By assumption, the heterozygote survives through its life better than the homozygous types and even produces more gametes and offspring. After all, that is what we mean by heterozygote superiority. But when it's time to reproduce, if it reproduces sexually, a heterozygous organism cannot breed true, that is, it cannot produce only heterozygotes like itself. Because of sex many of its offspring are the other homozygous types, *AA* and *aa*. Even when it mates with other heterozygotes, approximately a quarter of the offspring are *AA* and a quarter are *aa*. If the organism were to reproduce asexually, all the offspring of a heterozygous type *Aa* would be *Aa*, and the heterozygote would, as predicted by survival of the fittest, enjoy the greatest evolutionary success. But because of sex this does not happen.

Darwinian Dynamics

We have seen in Chapters 2 and 3 that natural selection depends on three characteristics: variation, heritability and what Darwin termed the "struggle to survive." We wish to develop a model that is general and not specific to a particular kind of organism or environment so that our conclusions will hopefully apply to any and all species (so long as Darwin's assumptions hold). Let us consider several different types of organisms, say, birds with different sizes of beaks, or giraffes with different-length necks. We use the subscript i to refer to these different types, for example, $i = 1$ might refer to a bird with a large beak and $i = 2$ to a bird with a small beak. Let N_i be the numbers of type i. Let W_i be the Darwinian fitness (the expected number of offspring produced by the ith type) and F_i the Fisherian fitness (the per capita rate of increase of the ith type).

For the moment, we ignore the complications of genetic variation within a species and let the subscript i denote different species of organisms, say different species of birds with different beak sizes, or different clones of, say, an asexual species of lizard or insect. We consider a situation where the different types compete for the same resource needed in reproduction, such as seeds in the case of birds or insects in the case of lizards. A simple representation of such competition would be $W_i = b_i R - d_i$, where R represents the resources available for reproduction, b_i is the number of offspring produced (if there were plenty of resources), and d_i is the probability of death. The birth and death parameters are different for each type i. (One way of thinking about available resources is to imagine how many resources there would be if there were no organisms around to eat them; let that number be K. Now assume that resources are converted directly into organisms, so that if there are a total of $N_1 + N_2 + N_3$ organisms, there are $K - N_1 - N_2 - N_3$ resources left that can be converted into new offspring. Available resources is then $R = K - N_1 - N_2 - N_3$.) If we start with N_i organisms of type i, the numbers after one generation should equal: new births, $b_i N_i R$, plus organisms that survive from the previous generation, $(1-d_i)N_i$. The change in numbers of type i is then

$$\frac{\Delta N_i}{N_i} = b_i R - d_i,$$

where Δ means change in one generation. I have divided both sides of the equation by N_i to give the per capita rate of increase, Fisherian fitness, as a function of the adaptive capacities and resources.

It is hard to image a simpler representation of Darwin's principles in mathematical form. The variation in the adaptive capacities of birth, b_i, and death, d_i, is subscripted by i. The "struggle to survive" is represented in terms of resources, R, that are needed to reproduce, with R being a decreasing function of the total numbers of organisms. So as numbers increase, so does the intensity of the "struggle." Heritability is represented by the fact that the offspring of type i

have the same capacities b_i and d_i as does the parent. We are obviously ignoring the complications of sex in this example. We will refer to equations like the one above as "Darwinian dynamics," because they so concretely represent Darwin's principles. Other examples are considered in the Chapter Notes.

Darwinian dynamics can be viewed as causal statements about natural selection. As with any equation, there is an equal sign; the left-hand side is equal to the right-hand side. We can view the right-hand side as being the cause and the left-hand side as the effect. The Fisherian fitness of type i is on the left-hand side. (In this example, since there is no sex, the two kinds of operational fitness, Fisherian and Darwinian, are the same.) On the right-hand side of the equation are the causal components of evolutionary success: heritable adaptive capacities, b_i and d_i, the environment, R, and the genetic and reproductive system. In the case of asexually reproducing organisms, the genetic system is so simple that you almost don't think it's there. But it is, because offspring have the same adaptive capacities as their parents.

Adaptationist Paradigm

A new type of organism often arises in a population by mutation (or migration from another population), and finds itself rare in its new home because most individuals are of a different type. When will the new type, we will call it "type 2," invade a population dominated by the resident type, which we will call "type 1"? By analyzing the Darwinian dynamic introduced above, we find that the answer to this question depends only on the adaptive capacities of each type of organism. Specifically, if $b_2/d_2 > b_1/d_1$, type 2 invades and eventually replaces type 1. The ratio of the adaptive capacities depends only on properties of the type of organism and, therefore, b_i/d_i is a measure of overall adaptedness of type i. Note that b_i/d_i is not the same thing as evolutionary success that is given by the left-hand side of the Darwinian dynamic equation above.

There is no tautology here. We have a causal and predictive hypothesis of an important aspect of evolutionary success, whether a new type can invade a population dominated by other types. In this case, the fittest, as measured by b_i/d_i, do indeed survive the best over time, so "survival of the fittest" is true. This is the adaptationist position: any increase in adaptedness, as measured by b_i/d_i, results in replacement by the new type. There is no limit to how rare that new type may be.

Darwin was, at times, an adaptationist. For example, he wrote in *The Origin of Species*, "under nature, the slightest difference of structure or constitution may well turn the nicely balanced scale in the struggle for life, and so be preserved" and that "each new form will tend in a fully stocked country to take the place of, and finally to exterminate, its own less-improved parent or other less-favored forms with which it comes into competition." The adaptive advantage may be extremely small, but as long as it exists, it will tip the balance and result in replacement of the old with the new. Only adaptedness matters: this is the bottom line for the adaptationist paradigm.

The validity of the survival-of-the-fittest paradigm depends critically on the kind of *density dependence* assumed in the Darwinian dynamic introduced above, and this in turn depends on whether the species is reproducing sexually. (There are other reasons the adaptationist paradigm is not robust and these are mentioned in the Chapter Notes.)

Density Dependence

We say there is density dependence when survival (and/or reproduction) depends on the numbers of organisms that are present in the population or community. In the Darwinian dynamic used above, there is no density dependence of death, as the capacity to die (d_i) is found alone on the right-hand side. Birth is density dependent, however, because the number of births is a product of the capacity at birth and available resources (b_iR). The available resources in turn depends on the numbers of organisms of both types 1 and 2 $(R = K - N_1 - N_2)$. This kind of density dependence results from the competition for resources and is shared by the two competing types. When the two competing types are different species, we speak of "interspecific" density dependence, when the density of one species affects the fitness of another species (as in the above case).

There are other kinds of density-dependent effects, termed "intra-specific" effects, that depend only on the numbers of organisms of the same species. For example, the chances of mating in a sexual species will usually depend on the numbers of potential mates that are present in a region. It is an easy matter to include these intraspecific density-dependent effects in Darwinian dynamics and this is done in the Chapter Notes.

When intraspecific density dependence is included in the birth and death components of the Darwinian dynamic, new and unexpected outcomes emerge. Survival of the fittest is only one of the possibilities. By analogy with survival of the fittest, we term these other outcomes "survival of the first" and "survival of anybody."

Survival of the First

Survival of the first refers to a situation when the new types cannot increase when rare even if they are more adapted than the resident type. It occurs when there is a cost to being rare. By being the first to colonize an area, a type will come to exist in large numbers and this prevents invasion by new types, even if the new types are more adapted. In short, no type, even one that is more adapted, can increase when rare. For this reason, we will also refer to this case as the "cost of rarity." The cost of rarity tends to maintain the status quo, or stasis, as an established community is buffered from invasion by other types.

A cost of rarity often occurs when organisms reproduce sexually. Sex requires mating and mating requires finding another member of the species. The likelihood of finding a mate depends on the numbers of mates available. For this reason, sexual reproduction is inherently density dependent. Recall from Chapter 5 that in facultatively sexual species (species in which both sexual and asexual individuals exist) the sexual forms tend to be produced when population density is high. When there are many members of a species in a given region, the costs of finding a mate should be less for each member. It is easier for the benefits of sex to outweigh the costs in dense populations, and for this reason I think sex is associated with high density. The cost of rarity will figure prominently in the story we have to tell about the origin of species in Chapter 10.

Survival of Anybody

"Survival of anybody" refers to a situation when a new rare type invades no matter what its adaptedness. It occurs when there is a cost to being common. Since the label "survival of anybody," if taken literally, sounds foolish and might obscure the interesting implications this case has for common populations, we will refer to this case as the "cost of commonness." The cost of commonness is a diversifying force. As already mentioned, parasites and predators tend to fixate on common types, freeing resources that allow other rare species to increase even if they are not particularly well adapted.

Darwin recognized the importance of the cost of commonness when he said, "When a species, owing to highly favorable circumstances, increases inordinately in numbers in a small tract, epidemics—at least, this seems generally to occur with our game animals—often ensue: and here we have a limiting check independent of the struggle for life."

Adaptedness Doesn't Exist

One consequence of the cost-of-rarity and cost-of-commonness paradigms is that in these situations there is no overall measure of adaptedness that predicts evolutionary success. "Survival of the fittest" is uninterpretable even though Darwin's principles of variation, heritability and the struggle to survive and reproduce are met. No overall property of the organism can be used to predict the evolutionary success of its genes. Darwinian dynamics, which faithfully represent Darwin's principles, often fail to support Darwin's and Spencer's central claim about the outcome of natural selection.

Yet evolutionary biology is full of thinking to the contrary. For example, the great population geneticist Theodosius Dobzhansky refers to adaptedness as the "*ability* of the organism to survive and reproduce in an environment." The evolutionary ecologist Eric Pianka calls it the "*conformity* between the organism and its environment." Philosophers have referred to overall adaptedness as a *propensity* existing in the organism to survive and reproduce. For example, the philosopher of biology Robert Brandon speaks of a "*biological property of an organism* in an environment measured by its expected fitness value." The problem with these concepts (emphasized by my italics in the above quotations) is that we cannot measure them in a nontautological way that reflects their use in evolutionary explanations.

No wonder the *Harper's* magazine author was confused. To listen to evolutionary biologists talk, one would think there was a property called adaptedness and that biologists had measured it. Bethell was correct to take biologists to task for their claims. But if we look at what evolutionary biologists do, not what they preach, we find their explanations to be rich in empirical content. No tautologies exist in the practice of evolutionary science.

Explicit mathematical models of evolution accurately reflect how evolutionary biologists explain nature. Darwinian dynamics express the rate of increase of a type as a function of the environment, the adaptive capacities, and the reproductive and genetic system. The adaptive capacities are likely candidates for measures of adaptedness: however, there is no general way of combining them into some overall measure (except in limiting situations like the case considered above). We cannot abstract an overall sum of adaptive capacities as an overall property of an organism. The only measure of a purported overall adaptedness property is Fisherian fitness, which leads us back to the tautology challenge.

No Parallelogram Law in Biology

In the Darwinian dynamic considered above, there were just two adaptive capacities. Typically, there are more adaptive capacities in the problems evolutionary biologists tackle. For example, a model of competition might also assume that the different types are more or less efficient at using the resource. This could be represented by type-specific carrying capacities, K_i, instead of the common capacity K. If we were interested in the evolution of helping behavior in a species of primate, we would define type-specific capacities involving the costs and benefits of the helping behavior. There is no end to the adaptive capacities we might consider, as our choice depends on the problem and question we wish to ask. Can these capacities be combined into an overall measure of adaptedness? Typically, they can't. There is no parallelogram law in biology for combining adaptive capacities as there is in Newtonian physics for combining force vectors.

There is, of course, the Darwinian dynamic itself, specifically its right-hand side, that combines the adaptive capacities into a predictive and causal statement of evolutionary change. The right-hand side of Darwinian dynamics combines adaptive capacities with the other components of natural selection—the environment and the genetic and reproductive system. But the right-hand side is not a property, ability, conformity, propensity or what have you, of the organism. It depends on everything. And this is the very point I wish to make. Natural selection does not just depend on "fitness," that is, the adaptive capacities of organisms.

Chance and Natural Selection

Survival of the fittest is not a tautology because fitness (the adaptive capacities of types of organisms) is just one component of survival (the evolutionary success of their underlying genes). Instead of illuminating the diverse components of the natural selection process, philosophers who have worked on the tautology problem have tended to focus on chance events while maintaining the fiction that adaptedness as an overall aggregate property of organisms exists. This isn't the place for a careful examination of their ideas, but in simple terms their argument runs as follows. The touchstone case they consider is that of identical twins. Identical twins must have the same "fitness," where fitness means an overall *propensity* in the organism to survive and reproduce. But one twin is struck by lightning and one isn't. The one that survives has the same fitness as the one who dies, so clearly survival of the fittest is not a tautology. Chance, so the argument goes, intervenes and saves the tautology.

Again, no wonder *Harper's* magazine was confused. If the empirical content of Darwin's great theory of natural selection rested only on chance, evolutionary biology would be in sad shape. Are we to believe that if the world was rigidly deterministic, with no chance effects, the theory of evolution by natural selection would be tautological? Of course not; the many components of the process of natural selection and the way in which these components are combined into causal explanations are what give natural selection empirical content, not chance.

Evolution does, of course, involve both chance and necessity, as we have discussed many times in this book. And chance can intervene to keep the fittest from surviving. But there is so much more to the process of natural selection, especially in sexual populations, that is ignored by focusing only on chance. I hope I have conveyed some of this complexity and richness.

Fitness is Not a Propensity

A sequence of tosses of a die makes a good analogy to the way in which chance enters into the actual process of natural selection. For example, say we were interested in obtaining a six. The actual observed frequency of six in a sequence of tosses of the die is analogous to operational fitness plus chance (we may get no six in any of our trials). The a priori probability of obtaining a six, calculated using physical principles and properties of the die (perhaps the dots on the die are magnetized and there is a strong magnet underneath the table), correspond to the functional dependence of Fisherian fitness on adaptive capacities in interaction with the environment and the genetic and reproductive systems. A bias in a die, say, its slightly deformed shape, or its having magnetized dots, are analogs of heritable adaptive capacities of an organism. The determination of which side of the die lands up by specifying relations between the properties of the die and conditions of the tossing (the environment) corresponds to functional determination of Fisherian fitness. It may be convenient to speak of the die as having a propensity to come up six, but I can't see why this helps understand what is going on.

Consider another example: the top speed of a newly designed automobile. The top speed attained will depend on the road, so let's consider a specified testing tract under highly controlled conditions. Obviously, the automobile has a top speed on a given testing tract; we drive the car and find it to be 120 miles per hour. So I suppose it makes sense to speak of the automobile as having a propensity to have a top speed of 120 miles per hour on that testing track. But this propensity can't be effectively determined independent of actually going out and driving the car on the track. The "speed propensity" of the car is just a way of talking about how various properties of the car interact with the road environment to determine the top speed. We certainly don't have an independent criterion for this propensity, but more importantly we don't *explain why* its top speed is 120 miles per hour on the track by appealing to an overall propensity to attain 120 miles per hour.

What the car has is gears, pistons, valves and tires all with specified capacities, that are combined together in a specified manner. A real smart mechanical engineer, familiar with the gear ratios and engine specifications might be able to combine all these properties of the car along with properties of the track into an equation that could predict a priori the top speed of the car on the track. This would be like having an equation of the form of the Darwinian dynamics considered above: top speed expressed as a function of capacities of the car along with the environment. The right-hand side of such an equation would not be a property of the car alone—like the Darwinian dynamic, it would combine all relevant parameters and variables into a causal and nontautological statement of expected top speed.

At any functional level a similar analysis could be conducted. Consider the gears, valves, pistons and tires that make up the car. They all have specified capacities that could be determined operationally or in a predictive and causal manner. For example, the force generated by the pistons and valves could be measured operationally, or they could be predicted using mechanical engineering principles and properties of the steel used, diameter of the valve, etc. Similarly, for the piston itself, and so on. The same infinitely nested set of capacities and properties holds for organisms. The adaptive capacities, b_i and d_i, are themselves functions of other capacities that could be specified should the question warrant it. For example, in a bird species the birth rate at unlimited resources may depend on mating success that may further depend on the attractiveness of the plumage, all of which could be expressed in mathematical form using type-specific parameters representing the capacity at mating or attractiveness.

We can speak of organisms as having a propensity to survive and reproduce, but I can't see how it helps us to understand what is going on. Furthermore, this propensity can't be defined independently of actual survival and reproduction. In fact, referring to such a propensity obscures how natural selection works, since an overall propensity confounds the causal components of natural selection that are better left explicit.

We have come to the picture of evolutionary explanations shown in Figure 1. All causal systematic components of natural selection are reflected in the Fisherian fitness (*F*-fitness), the per capita rate of increase of the type. *F*-fitness is the bottom line concerning natural selection. Evolution, however, also involves chance effects.

Biology is Fitness; All Else is Chemistry and Physics

Fitness is *the* uniquely biological concept. All else is molecular biology, chemistry, and physics. The theory of evolution by natural selection should explain fitness. As we have found in this chapter, we should not look for explanations of fitness as an overall or aggregate property of the organism. To do so is to confound and blur its causal components. In addition, use of this concept in evolutionary

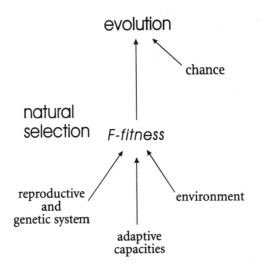

Figure 1. Fitness and evolution

explanations is tautological. The fitness we find in explicit mathematical models of natural selection is Fisherian fitness, F-fitness, the per capita rate of increase of a type. To explain F-fitness, we look to causal relations between adaptive capacities, environmental variables, and genetic and reproductive systems. In these relations we find hypotheses of particular adaptive capacities and how they causally contribute to F-fitness. How, then, do we explain the present adaptiveness of adaptive capacities? Because of the repetitiveness of life cycles, effects in one generation become causes in the next generation. Adaptive traits are both a cause of F-fitness differences in the present and an effect of F-fitness differences in the past.

Organisms Don't Evolve

It is easy to be confused by organisms and their role in evolution. Endowed and ornamented as they are with such wondrous designs, it is easy to see them as the product of evolution. We are organisms and we seek to understand organisms to understand ourselves. But organisms are of only fleeting existence. Each is born genetically unique (because of sex) and each lives but a moment in evolutionary time. Organisms don't evolve. When viewed from the perspective of the evolutionary process, organisms appear as vehicles for the perpetuation of genes. Even "fitness" is best understood as a process of genetic change rather than as a property of organisms. Evolution is all process: it has no ultimate products.

Darwin's Dilemmas

> *Why is not all nature in confusion instead of the species being, as we see, well defined? ... Why are not all organic beings blended together in an inextricable chaos?*
> —Charles Darwin,
> The Origin of Species

The world is an orderly place, at least for the time being. The laws of physics tell us that the universe is becoming more and more disordered, that all structure must decay and eventually disappear. Nevertheless, the relentless push toward chaos and confusion can be delayed, at least locally, on planets like our own, planets that are bathed in entropy-procrastinating sunshine. Sunshine is the source of all biological order; sunshine fueled the transition from chemical order to biological order some 4 billion years ago and continues to be the energy that maintains life on our plant today.

Order from Chaos

While sunlight is the fuel supporting life, natural selection is the engine that moves life forward. To channel the sun's energy into useful structures requires its careful guidance. Without natural selection life would soon coast to a stop. As we found in Chapter 2, natural selection itself emerged from the chaos, when replicating molecules occurred spontaneously 4 billion years ago. The useful structures created by natural selection adapt organisms to new environments, enabling them to reproduce and pass on their genes. Organisms are impressive because of their adaptations, and these beautiful designs are one aspect of biological order.

Species are a second aspect of biological order. The living world is not a blend of living things, each slightly different from the next. Rather the living world is organized into the categories we call species. Why do species exist? This question is the question Darwin really wanted to solve, as he announced to the world in 1859 in the title of his great book, *The Origin of Species*. For Darwin, the question of the origin of species was the same as "Why are not all organic beings blended together in an inextricable chaos?" There is order in the living world; it is not an inextricable chaos. Species are, for the most part, distinct and recognizable; we give them names. Cultures without the formal study of modern biology name the same species as do trained biologists. True, there is confusion within certain groups of organisms, (for example, in some groups of plants) but species are generally distinct and different from one another. They are so distinct that early biologists took them as evidence for the unchanging categories created by God.

Why Do Species Exist?

In the nonliving world, the word "species" means kinds of, as in different species of molecules or minerals. Nonliving species are defined in terms of a list of jointly sufficient and individually necessary characteristics that determine whether an object is a member of a certain class. A casual look at biology might give the impression that biologists, too, use the word "species" in this sense. For example, popular books on animal or plant identification give lists of properties that allow the amateur naturalist to determine whether the bird outside their window is a sparrow or a finch. Indeed, in nature organisms do share many characteristics with members of their own species and far fewer characteristics with members of different species. However, biological species are also defined by the gaps that seperate them.

Species membership is usually not a matter of degree but of discontinuity. No intermediate types of organisms typically exist with characteristics in between the characteristics typical of two good species—especially when these species are similar and found together. The gaps that separate different species are as basic to the biological species concept as are the characteristics that are shared by members of the same species.

Early ecologists such as Charles Elton and David Lack argued that morphologically similar species could not coexist. Population biologist Evelyn Hutchinson even suggested a minimum size ratio of about 1.3 to 1 between members of similar coexisting species. When comparing competing species, it is important to consider species from the same natural group of competitors, and to consider traits that make sense in terms of the species' natural history. An especially well studied example concerns size of canine teeth in North American weasels, mustelids, and small cats in the Middle East. The species studied tend to share prey, and to kill their prey in a similar way by biting with canines. Canine diameter in these species is discountinuously distributed with evenly spaced gaps, while another morphological character not related to feeding, skull size, tends to be continuously distributed.

For early biologists, like the great eighteenth-century Swedish botanist Carl Linnaeus, as for the fundamental biblical creationist, there is no problem with species discontinuities. The apparently bridgeless gaps that separate species are modern evidence of God's ancient handiwork. The gaps in existence today were created by God and preserved for all time by the fidelity of reproduction, so thought seventeenth- and eighteenth-century biologists. The essence, or *edios*, of each species in Plato's ontology was thought to somehow be contained in the seed and thereby passed from parent to offspring through the eons. In this view, species have not changed since the Creation. But even Linnaeus eventually admitted that species do in fact change; some species had even gone extinct. How can one reconcile the changeability of species with gaps? The apparently bridgeless gaps must have been bridged. But when and how?

Missing Links

The problem of bridging gaps becomes especially acute if evolution, as Darwin argued, is a continuous process proceeding in small steps.

What has happened to the intermediate steps that must connect all living things, even members of different species? This problem concerned Darwin greatly. In his own words: "As according to the theory of natural selection an interminable number of intermediate forms must have existed, linking together all the species in each group by gradations as fine as are our existing varieties, it may be asked; Why do we not see these linking forms all around us? . . . Why, if species have descended from other species by insensibly fine gradations, do we not see innumerable transitional forms?" Darwin's problem of missing links, as we will call it, is actually two separate but related problems: missing links in habitat space and missing links in time.

Missing links in habitat space are the kind of thing we are talking about when we go outside, look around, and can't find an intermediate type, say a bird with beak size in between two species of finch. We might imagine, as did Darwin, that the intermediate types, the links, existed in the past but have gone extinct. We look for their fossils in the fossil record, and often we can't find any linking types there either. This is the problem of missing links in time.

Darwin's Solution to Missing Links

Darwin expected that natural selection should produce "innumerable transitional links" if evolution occurred in "an extensive and continuous area with graduated physical conditions." He further argued that intermediate types are not observed among current living forms because the intermediate regions between two species are small. Small regions means small population sizes, and he believed that small populations evolve more slowly than large populations and therefore would be eliminated in the long run by competition with the more rapidly evolving species. He wrote in *The Origin of Species*:

> The neutral territory between the two representative species is generally narrow in comparison with the territory proper to each. The intermediate variety, consequently, will exist in lesser numbers and during the process of further modification through natural selection, they will almost certainly be beaten and supplanted by the forms which they connect; for these from existing in greater numbers will,

in the aggregate, present more variation, and thus be further improved through natural selection and gain further advantages.

Darwin realized that his solution to the dilemma of missing links in habitat space created his second dilemma of missing links in time since "just in proportion as this process of extermination has acted on an enormous scale, so must the number of intermediate varieties, which have formerly existed, be truly enormous." As is well known, to resolve the second dilemma he appealed to what he viewed as "the extreme imperfection of the geological record."

Unfortunately, evolutionary biology has all but ignored Darwin's dilemmas. No satisfactory explanation of gaps and species' distinctness can be found in the great treatises of Mayr or Simpson that deal with almost every other aspect of species. Darwin's dilemmas continue to be a problem in understanding the distribution of genetic and phenotypic variation in time and space. One perspective on missing links in time is based on the hypothesis of punctuated equilibria in which evolution consists of periods of quiet status quo (stasis) punctuated by relatively rapid changes in species and community compositions. As we know from the last chapter, sex can produce stasis through its cost of rarity which tends to stabilize existing communities. Current theories on the question of missing links in habitat space are based on the ideas of "character displacement" and "limiting similarity." A complete discussion of the effects of sex, and the concomitant costs of rarity, on these issues is beyond the scope of this book; but the relevant literature can be found in the chapter notes.

Sex Has a Cost of Rarity

I believe the resolution of Darwin's dilemmas lies in sex and the fact, introduced in previous chapters, that sexual reproduction carries an intrinsic cost of rarity. This cost of rarity tends to create and maintain species as distinct and broadly homogeneous groupings of individuals. We will think of the cost of rarity as the cost of finding a mate, although this is an oversimplification of many related issues.

Organisms that have adapted to rarity may not appear to have problems finding mates as population numbers decline, but that is because they devote energy, time and genes to structures that enable them to successfully find mates. For example, plants allocate more energy and resources to flowers and mating as plant or pollinator density decreases. This energy is no longer available for the plant's maintenance and survival. As a result, the cost of finding a mate may not be apparent as a decrease in mating success but rather a decrease in survival.

Even in our own technologically advanced species, the chances of finding a mate depends on population size. Farmers and ranchers who live in sparsely populated areas expend considerable time, energy and money finding a mate. Special publications even exist to aid the process of finding and selecting a mate in these communities. The basic issues involved are easy to appreciate and simple to represent in abstract terms. We do so now that we may understand the effect of sex on species.

Modeling Sex and Species

Consider two sexually reproducing species named species 1 and species 2 with population sizes N_1 and N_2. To make matters as simple as possible, we ignore any effect of density on mortality and assume that population size affects only birth. (That is we take $u = 1$ and $v = 0$ in the equations presented in the Notes to the last chapter.) We assume that the chances of finding a mate are proportional to population density (numbers of organisms per unit area). Consequently, for each sexual species, 1 and 2, the realized number of births depends on the numbers of each species, N_1 and N_2, as well as on the intrinsic adaptive capacities at birth, b_1 and b_2, and resources, R. The following Darwinian dynamics then apply.

$$\frac{\Delta N_1}{N_1} = b_1 R N_1 - d_1$$

$$\frac{\Delta N_2}{N_2} = b_2 R N_2 - d_2$$

$$R = K - N_1 - N_2$$

Even if the two species were identical ($b_1 = b_2$, $d_1 = d_2$), the resident species would enjoy a higher *per capita* rate of increase because of its larger numbers. In other words, a member of the common species has greater reproductive success than a member of the rare species, simply because its mates are more common.

Mating alters the basic form of the Darwinian dynamic by introducing nonlinear terms, such as N_i^2, into the birth part of the equation. The absolute numbers of new births depends on N_i^2, instead of just N_i as it does for an asexual species (to see this, multiply each side of the above equation by N_i to get rid of the $1/N_i$ term on the left-hand side). The cost of rarity for sexual species stems from this nonlinearity, because it accentuates the effect population size has on the rate of increase of each species.

Speciation on an Environmental Continuum

Many of the basic environmental variables upon which life depends, such as temperature or moisture, usually change continuously as a function of position, a situation we will refer to as an "environmental continuum." Although some variables change discontinuously—for example, at the surface of the ocean where moisture changes dramatically in and out of the water—it is instructive to consider an environmental continuum as a worst case. Any process that produces species distinctness on a continuum should only be enhanced by a distinctness in the environment. Indeed, Darwin saw the need to consider this situation of evolution occurring on "an extensive and continuous area with graduated physical conditions."

As species adapt to their environment, they tend to specialize on specific environmental parameters and conditions. For example, a species of fish may live within a narrow range of water temperature, or a bird species may be found only at a certain altitude up a mountain. As the biota continues to diversify, new species are formed, each adapted to a more limited region of the environment. Consequently, more and more species come to be packed into the same region of the parameter space (the mountain or range of temperatures). The key assumption in my argument is that as more and more species occupy a continuous region of the environment, numbers of organisms in each species get smaller—that is, the species get rarer. The reason is that we are considering species that are generally similar, and the environment can only support a certain number of organisms that are of the same general kind. As the species become rarer, the costs of finding a mate increase for each member of the species. The problem for sexual species reduces to considering whether the advantages of adapting to more local conditions can overcome these costs of becoming rarer.

In Figure 1, the fitness (per capita rate of increase) of a species is given as a function of a continuous environmental variable, like temperature or altitude. We assume that each species does best at some particular point along the environmental continuum. For example, each species of plant has an altitude at which it does best. This optimal value of the environmental variable is represented by a peak

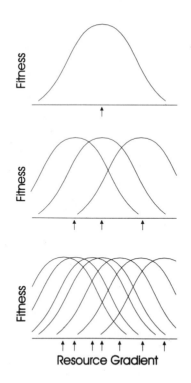

Figure 1. Fitness and speciation

in the fitness curve in the figure (an arrow shows the optimum). Each curve represents a different species. If several species are present, the distance between the peaks reflects the differences between the species. For similar species the peaks are close together, while for different species the peaks are far apart. Species are distinct to the extent that there are gaps between the peaks so that the curves do not overlap.

In the middle panel of the figure, we show the case in which the original species has split off two species, one to each side. Finally, in the bottom panel, we show the case in which each of the two daughter species in the middle panel have themselves split off two more species. It is easy to see that the differences among the fitnesses, and hence the benefit of further splittings, decrease with the number of splittings and number of species.

Costs of Splitting

Let us now turn to the costs of splitting. As the number of species occupying a region of the environment increases, the numbers of organisms in each species will decrease, for reasons we have just discussed. In the simplest case, if the number of species doubles, the number of individuals in each species decreases by about one-half. Consequently, every individual will experience increased difficulty in finding a mate, since there are now fewer potential mates in each species. One can imagine two responses, each one of which leads to the same result. In one case, a species does nothing about this increased difficulty in finding a mate, and the cost of splitting is exacted in terms of fewer matings, and, consequently, fewer offspring. In the other case, a species might increase the resources devoted to finding a mate (as did the plants mentioned above) at the expense of some other critical function like maintenance or survival. In that case, the cost of finding a mate is exacted through the reduced contribution of this other need to fitness. In either case, each splitting event adds a fixed amount to the cost of further splittings, so the cost of splitting also increases with the number of splittings and number of species.

This cost-benefit analysis of species splitting (speciation) is a straightforward example of the law of diminishing returns for sexual species. As the number of splittings increases, the benefits of further splittings get smaller and the costs get larger. At some point, the costs of further splitting outweigh the benefits. At this point, the partitioning of the habitat must stop, the distribution of sexual species is stable, and species distinctness has been achieved. The resulting gaps between species is the limit to their similarity, and these gaps result from the cost of rarity inherent in sexual reproduction.

Order Out of Chaos

If nature begins in confusion, can order result? For reasons similar to the ones just discussed, a continuum of sexual species, each slightly but imperceptibly different, is unstable. Because of the cost of rarity, it will coalesce into a discrete distribution of distinct species. Let me explain why.

We simplify matters by taking the initial distribution of species to consist of equal numbers of extremely finely-adapted species—as in the bottom panel of Figure 1, but with even more species so that they merge together in a blur. Assume that by chance the number of organisms in one species increases and the number in another species decreases (so the net change is zero). Assume also that there is no affect of these changes on fitness, because the resulting changes in resource use and levels of resources are so small that they can be ignored (this is reasonable if the niches as characterized by the peaks in the fitness curves of these two species are contiguous). In the case of a continuum of species, each species is very rare to begin with. Because of this initial rarity, the influence of small changes in numbers of potential mates give rise to large changes in the likelihood of finding suitable mates. This causes the fitness of the species with the increased population to be larger relative both to the species that it partially displaces and also to the others in the region whose population numbers remain the same. The species with the small increase in numbers then starts increasing further and the others start decreasing. As it does, its own cost of finding a mate decreases while the cost of finding a mate for all the other species continues to increase. There is thus a snowballing effect in the instability, insofar as it proceeds with increasing speed as it develops. The snowballing instability keeps rolling, gaining momentum, until all species in nearby niches are eliminated. The species that initially occupied a very narrow region of the environmental grade now occupies a considerably larger niche, nearby species have gone extinct and distinctness of species has been achieved. Order can result from chaos in sexually reproducing species because of the cost of rarity inherent in sexual reproduction.

Cost of Rarity Creates Species

Numerical studies using computer experiments of mathematical models of competing species support these verbal arguments. In Figure 2 is an example from one numerical calculation to illustrate the

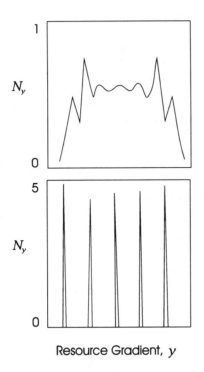

Figure 2. Species distinctness and sex

points just made. The plots themselves represent the stable equilibrium populations of species along a continuous resource grade, denoted as y. The width of the resource grade is about six times wider than the width of any one fitness curve. In this case, no one species can utilize all the resources, so several species are expected. In the top panel, we show the outcome of competition between species for which there is no cost of rarity, such as for a group of asexual species. The computer predicts a dense set of vertical lines whose heights give the sizes of the populations. The lines are too dense to be plotted; instead we show a curve drawn through the tops of the lines. Nature is in confusion: each species is only slightly distinguishable from its neighbor.

In the bottom panel, we show the outcome of the case in which each species has an intrinsic cost of rarity. Most species have been eliminated by competition, leaving 5 species to occupy the resource grade. Comparing the two panels, we see a change from an indefinite proliferation of types for asexual species to a limited set of discrete sexual species with gaps in between.

Asexual World in Confusion

For asexual organisms, there is no intrinsic cost of rarity and only benefits to fine-scale adaptation to the environment. Linking species should be observed in nature, as the top panel depicting the computer experiments indicates. There should be no missing links and no dilemmas of missing-links. Asexual species should exist in confusion and proliferate into a continuous distribution of finely adapted types. Data from asexual clones support these expectations, since "clonal swarms" of slightly different types are commonly found, with rare exceptions, in obligate parthenogenetic animals (animals that reproduce by development of an unfertilized gamete).

The parthenogenetic weevil (*Otiorrhynchus scaber*) is a common forest pest in Finland and Sweden and has been extensively studied for this reason. Its geographic distribution is typical of asexual species. In one study, 4 individuals were sampled from each of 123 geographic locations in Scandinavia. In Sweden, 214 individuals were studied genetically and were found to be made up of 29 clonal genotypes; in Finland, the 261 individuals studied comprised 66 clones. The different clonal genotypes were shown to have slightly different ecological requirements. Bearing in mind that this is a very limited sample of an abundant population of forest pests, one can be certain that there are vastly more genotypic clones in the population than have been found so far. Slightly different reproductively isolated clones coexist in small regions of space, something that could not happen with reproductively isolated sexual species.

Let us compare this clonal swarm with a sexual example taken from another insect, the mosquito genus *Anopheles* in Europe. This population consists of 6 "sibling" species, that is, 6 morphologically similar sexually reproducing species. We choose sibling species to allow us to compare two sets (asexual and sexual) of morphologically similar organisms, both of which are subdivided into reproductively isolated and potentially sympatric subunits. In contrast to the clonal weevils, these mosquitoes have distinct ranges and habitats. Two are from northern Europe and breed in fresh and brackish water, respectively; two others are in southern Europe and are likewise divided by the salinity of water. One lives in mountains and another is limited to the Middle East. The populations are largely geographically disjunct, and where they are sympatric there are well-defined differences in the type of water used for breeding. On the scale of uniformity of habitat occupied by the clones, these morphologically similar species are distinct geographically or in habitat use.

The difference between these two cases is quite stark. In the sexual case, there are only six populations, and these are seen to segregate in habitats on a coarse scale involving large and distinct geographic areas, different breeding preferences, different conditions of temperature, humidity, etc. In the clonal case, some of the clonal populations segregate in comprehensible ways relative to some of the others: however, the population as a whole consists of a profusion of types all intermixed. It is clear that even the most skilled naturalists would be challenged to identify the relative habitat preferences of all of the clones.

The clonal populations of weevils are typical of our expectation for the asexual world of nature in confusion: a proliferation of types, each adapted to the finest imaginable nuance of the environment, limited only by the criterion that each population must be represented by at least one individual. In the sexual example of mosquitoes, we find a coalescing of organisms into a limited number of basic units—species—each occupying a coarse slice of the environment and separated from other species by gaps in resource and environment use.

Unity and Distinctness Through Sex

Sex has always been seen as fundamental to the biological species concept. For example, one of the world's most distinguished authorities on species, Ernst Mayr, defines species as groups of interbreeding, or potentially interbreeding, organisms. The intermixing of genes within a sexual species results in a reticulated pattern of interconnectedness and unity that is characteristic of sexual biology. It is this web of shared information that makes all living things truly one. The "unity of the genotype," as Ernst Mayr calls it, refers to the cohesive nature of gene systems in sexual organisms. However, the unity of species is not the same thing as the distinctness of species. There is nothing in the unity of the genotype that requires gaps between species. Highly complex and cohesive systems can be slightly different from one another. (One needs only consider the slew of computer word processing programs on the market today to be convinced of that!)

Species distinctness refers to the gaps in gene and phenotype space that often distinguish one web of interconnectedness, one species, from another. But, as we have found, sex is fundamental here, too. The nonlinearities inherent in the dynamics of sex produce distinctness of species. Sex has two seemingly opposite, effects on the living world: unity and distinctness.

Sexual Biology

Interconnectedness and unity among distinctly different (yet still interconnected) groups of organisms. This is sexual biology. A sexual biology contains unique organisms that exist but fleetingly as temporary repositories for gene combinations, each combination existing for just a blink of evolutionary time until it is destroyed by the same sexual process that created it. Organisms are born and they die: in between being born and dying they have offspring, some more than others according to the traits they possess. Their wondrous adaptations are impressive. However, because of sex, organisms as individuals play no enduring role in the evolutionary process.

Genes, however, go on forever. This is what sex is about. The interwoveness and timelessness of gene lineages. The immortality of life resides in the well-being of genes, which requires damage repair and the cleansing of mutations. Sex began as a necessary and mechanical process of gene repair. And it forever changed the landscape of the living world.

Chapter Notes

Chapter 1: Why Sex?

REFERENCES FOR MAIN TEXT

The topics introduced in this chapter are considered in greater detail in later chapters, and the relevant references can be found in the notes for the respective chapters. There are several excellent reviews of the problem of why sex evolved (Williams, 1975; Maynard Smith, 1978b; Bell, 1982). Recent work can be found in two edited volumes (Michod and Levin, 1988; Stearns, 1987) and the proceedings of a recent symposium (published in *J. Heredity,* vol. 84, 1994). Other material referred to in the chapter includes Weismann (1889) and Darwin (1859). A history of evolutionary thinking is given in Bowler (1984). The role of Wallace in Darwin's thinking concerning survival of the fittest can be found in Marchant (1975, 140–145). For a discussion of the relevance of Darwin's principles to order in physical and chemical systems, see Bernstein et al. (1983).

I use recombination to mean breakage and reunion of DNA molecules. This is the more modern usage common in molecular genetics. More classically, "recombination" means new combinations of alleles at different loci. I explicitly refer to new combinations of genes when this is what I intend. I use the words "crossing over" to refer to new combinations of genes at linked loci. Use of the word recombination can be confusing, because new combinations of genes and crossing over can, but need not, result from breakage and reunion of homologous DNA molecules (as explained in Chapter 6).

Chapter 2: The Early Replicators

REFERENCES FOR MAIN TEXT

The Garrett Hardin reference is Hardin (1968). The origin of life and genetic systems is a large field with diverse points of view (Oparin, 1938; Miller and Orgel, 1974; Bernal, 1967; Eigen et al., 1981; Oparin, 1965; Woese, 1980; Attwood, 1980; Cairns-Smith, 1982; Michod, 1983). The replicator point of view used in the chapter is discussed further in Eigen and Schuster (1979), Schuster (1980), Eigen et al. (1981), Eigen (1992), and Michod (1983). Enzyme-free template-based polymerization of oligonucleotides has been reported by Orgel and colleagues (Lohrmann and Orgel, 1980; Lohrmann et al. 1980; Van Roode and Orgel, 1980). The famous origin of life experiment mentioned in the text was done by Stanley Miller in 1954 (see Miller and Orgel, 1974). An introduction to entropy in the context of biological systems can be found in Morowitz (1970). One idea for the origin of the genetic code is that proteins may initially have been produced by direct pairing between the nucleotide base sequence and the amino acids (Hendry et al., 1981b; Hendry et al. 1981a; Woese, 1967). The evolution of the cell and the origin of the genotype and phenotype is further discussed in Michod (1983). My colleagues and I have considered the origin of sex from the point of view of genetic error in several papers (Bernstein et al., 1984b; Long and Michod, 1994; Michod and Long, 1994; Michod et al., 1989). The efficiency of haploid replication in comparison to diploid replication has been discussed in Lewis (1985), Destombe et al. (1993), and Valero et al. (1993). The competition of haploid, diploid and sexual cells has been studied using mathematical models (Long and Michod, 1994; Michod and Long, 1994). The theory that sex originated for the benefit of infectious elements has been developed by Donal Hickey and Michael Rose (Hickey and Rose, 1988; Hickey, 1982; Hickey, 1993; Rose, 1983). The idea that sex originated for the purpose of nutrition is discussed in several articles (Stewart and Carlson, 1986; Redfield, 1993; Redfield, 1994).

Chapter 3: Mrs. Anderson's Baby

REFERENCES FOR MAIN TEXT

Dr. Pangloss' observation in Voltaire's *Candide* that humans have noses so as to wear glasses was introduced to evolutionary biology by Gould and Lewontin (1979). An unsolved problem in sexual biology is the origin of isogamous mating types from a sexual population in which any cell can fuse with any other cell (Hoekstra, 1987). See Hoekstra (1987) for an overview of the evolution of anisogamy. An overview of the many costs of sex is given in Lewis (1987). The handicap principle and the runaway process have been extensively discussed in the literature (for example, Andersson, 1982; Andersson, 1986; Arnold, 1985; Bell, 1978; Davis and O'Donald, 1976; Eshel, 1978; Kirkpatrick, 1982; Kirkpatrick, 1986a; Kodric Brown and Brown, 1984; Lande, 1981; Maynard Smith, 1976a; Maynard Smith, 1978b; Maynard Smith, 1985b; Nur and Hasson, 1984; Pomiankowski, 1987a; Pomiankowski, 1987b; Trivers, 1972; Zahavi, 1975; Zahavi, 1977; Hasson, 1989; Kirkpatrick, 1986b; Maynard Smith, 1985a; Michod and Hasson, 1990). The runaway process was first proposed by Fisher (Fisher, 1930; Fisher, 1915; Fisher, 1958). The handicap hypothesis was first proposed by Zahavi (Zahavi, 1975; Zahavi, 1977; Zahavi, 1978). The figure in the honesty in advertisement section in the text is based on the work of Nur and Hasson (1984) and Michod and Hasson (1990). Cooperation and altruism can evolve by natural selection. The evolution of cooperation and altruism in populations in which relatives interact was first discussed at length by Bill Hamilton (Hamilton, 1963; Hamilton, 1964a; Hamilton, 1964b). Kin selection, as it is often called, has been extensively studied and a review of its theoretical aspects can be found in Michod (1982). The role of group structure and population structure in the evolution of cooperation has been considered by a variety of authors (Maynard Smith, 1964; Uyenoyama and Feldman, 1984; Wade, 1978; Wilson, 1975; Wilson, 1980). The reciprocation of cooperation is often favored by natural selection when organisms interact repeatedly (Trivers, 1971; Brown et al., 1982; Michod and Sanderson, 1985; Axelrod, 1984; Axelrod and Hamilton, 1981; Boyd and Lorberbaum, 1987; Lombardo, 1985; Nowak and Sigmund, 1992)—even when these organisms are highly mobile (Ferriere and Michod, 1994a; Ferriere and Michod, 1994b).

Chapter 4: Sex and Death

REFERENCES FOR MAIN TEXT

An overview of the effects of sex and reproduction on death and mortality can be found in the book by Stearns (1992). A grand-childless mutant of the common fruit fly *Drosophila* lacks ovaries and lives longer than normal females (Maynard Smith, 1959). Even sperm, often ridiculed because it is cheap and poorly constructed, is costly in simple parasitic worms (Van Voorhies, 1992). The evolutionary theory of aging based on pleiotropy and the decline of the force of selection with age was first described by George Williams and Bill Hamilton (Williams, 1957; Hamilton, 1966;). The role of DNA damage in aging is summarized in Bernstein and Bernstein (1991) and Bernstein (1979). The aging of single-celled protists and the effect of Muller's ratchet is discussed in Bell (1988). Concerning Freud's views on the origin of sex and psychological instincts, see especially Freud (1961).

Chapter 5: The Undoing of Sex

REFERENCES FOR MAIN TEXT

Throughout the chapter reference is made to the following books: Bell (1982); Maynard Smith (1978b); Williams (1975). Muller's ratchet was first described in Muller (1964). The role of synergistic selection in the evolution of sex has been discussed primarily by Kondrashov (Kondrashov and Crow, 1991; Kondrashov, 1988; Kondrashov, 1985; Kondrashov, 1984; Kondrashov, 1982; see also Redfield, 1988 for an application to transformation). The Fisher-Muller theory was first described in Fisher's classic book (1958; 1930) and in Muller (1932). The lottery metaphor was developed by G.C. Williams (Williams, 1975; Williams and Mitton, 1973). The role of sibling competition in Williams's lottery metaphor was considered by Maynard Smith (1977; 1976b). Bell showed that sex in facultative organisms is associated with high population densities (1988). Group selection on sex is discussed in several books and articles (Maynard Smith, 1978a; Nunney, 1989; Stanley, 1976). The importance of capricious environments and flip-flopping of environmental states has been considered by several workers (Charlesworth, 1976; Felsenstein, 1988; Maynard Smith, 1988; Maynard Smith, 1978b; Sasaki and Iwaas, 1987). The parasite theory for sex can be found in the work of W.D. Hamilton (Hamilton, 1980; Seger and Hamilton, 1988; Hamilton and Zuk, 1982; Hamilton et al., 1980; Hamilton, 1993). The level of linkage disequilibrium in natural populations of outcrossing organisms for randomly selected genes has been measured by the technique of gel electrophoresis in *Drosophila* and found to be small (Langley et al., 1974; Langley, 1977; Hedrick, 1983; and Hedrick et al., 1978). The quote from Maynard Smith at the end of the chapter comes from his book (Maynard Smith, 1978b).

INTERMEDIATE OPTIMUM MODEL

In the text, we consider parasites as a possible way to generate flip-flopping in environmental states. Another way in which the required flip-flopping might occur has been studied by John Maynard Smith (Maynard Smith, 1980; but see Bergman and Feldman, 1990) and involves selection for intermediate values of traits determined by many genes. Many traits, like height and weight, are affected by many genes, and often such traits have an intermediate optimum value: for example, being either too heavy or too light is not a good thing. Maynard Smith showed that if the optimum value changes from time to time this might generate a flip-flop in the desired association between the genes affecting the trait.

It is not too difficult to see why selection favoring an intermediate value of a trait will create associations between the underlying genes. Assume for explicitness again just two genes with two states each, say A_1 or A_2 and B_1 or B_2, affecting a trait like weight for which there is an intermediate optimum. Assume further that the *1* genes contribute nothing to the trait while the *2* genes contribute a positive value of unity to the trait. For example, an individual whose diploid genotype is $A_1A_1B_1B_1$ has weight zero while an $A_2A_2B_2B_2$ individual has weight four. Of course, it is silly to talk about an individual with weight zero, but we could be measuring the effects of the genes in terms of a deviation from some standard value, so that the individuals just discussed have a weight either equal to the standard value or four units above it. The important matter is that extreme individuals are assumed to have low fitness. It is better then to have a mixture of 1 and 2 genes, like individuals of genotype $A_1A_2B_1B_2$, who have an intermediate weight of two and as a result have the highest fitness. As a result of selection for intermediate values of such a trait, individuals with a mixture of *1* and *2* genes will be common in the population and individuals with many *1* genes or many *2* genes will be relatively rare. This will produce an association between the genes at the A locus and the genes at the B locus and it is just such an association that is required for sex to have an effect.

However, for sex to be advantageous we need a flip-flop of the favored association between the genes at the A locus and the B locus. Maynard Smith then assumed that the optimum value might change. He showed that constantly shifting the intermediate optimum value for the trait could create such a condition. Consider an initial state of the population in which A_1B_2 and A_2B_1 combinations are relatively common and A_1B_1 and A_2B_2 gene combinations are relatively rare. If the favored value of the trait changes, those combinations that were initially not favored and rare will become favored. Say the favored value of the trait increases: there will now be fewer A_2B_2 gametes than selection would favor. These combinations can be produced only by crossing-over; so sex is advantageous. As long as the optimum value of the trait continues to shift among intermediate values, a sex gene can increase in frequency. Subsequent mathematical work by Bergman and Feldman (1990) has raised some problems with Maynard Smith's hypothesis that we will not go into here.

LINKAGE DISEQUILIBRIUM AND SEX

To rigorously evaluate the theories that assume that the benefit of sex is genetic variability we need a measure of association in the population between alleles at two loci. Let A refer to one locus and B to the other. Let us consider just two genes with two alleles each, say A_1, A_2 and B_1, B_2. One measure of the association between the two traits is simply $P(A_1, B_1)$, where $P(A_1, B_1)$ means the probability of genes A_1 and B_1 being found together in the same gamete or on the same chromosome, and likewise for the other three combinations $P(A_2, B_2)$, $P(A_1, B_2)$, and $P(A_2, B_1)$. But we also need a measure that considers all the associations together so that we have some idea of the overall association between the two kinds of traits. One such measure that has been extensively used in population genetics (see, for example, population genetics texts by Crow and Kimura, 1970; Roughgarden, 1979) is the *coefficient of linkage disequilibrium*, $D = P(A_1, B_1)P(A_2, B_2) - P(A_1, B_2)P(A_2, B_1)$. It can be seen that D is zero if the genes for the traits are independent of one another and combine randomly, or, in other words, if $P(A_i, B_j) = P(A_i)P(B_j)$. In this case, there is no association between the genes for the two traits.

Nonrandom associations between genes have an effect on selection. If $D = 0$, then natural selection can go on independently for the two loci. Selection for adaptation to temperature, say, will not interfere with selection for adaptation to moisture. If the genes for the two traits are associated with each other, then selection for one trait will affect selection for the other. For example, selection for increased tolerance of high temperatures may drag along genes for increased sensitivity to moisture. The main effect of sex—indeed, this is the only effect of sex on population genetic variability—is to break down these associations and to allow selection on a trait to occur independently of other traits.

SEX DECREASES LINKAGE DISEQUILIBRIUM

Genetic associations and their removal by recombination are the common denominator of all theories of sex based on genetic variability. Sex can have an effect only on the distribution of traits in a population, and hence on the evolution of the population, if D is nonzero, that is, if traits are associated nonrandomly. Furthermore, if D is nonzero, sex has the tendency of reducing D to zero by making the distribution of traits more and more random. Sex destroys the very associations that are necessary for it to have an effect.

Decreasing linkage disequilibrium, D, is the fundamental consequence of sex—indeed, it is the only consequence of sex—on the combinations of genes found in populations. As a result, for sex to be continuously advantageous there must be some source that restores D to a nonzero value or else sex would take the population to zero D and then no longer have any effect. Through crossing-over, sex mixes traits from different individuals. If the population is already completely mixed ($D = 0$), further mixing can have no effect.

CAUSES OF LINKAGE DISEQUILIBRIUM

If sex can have an effect only in populations in which D is not zero, then all the theories discussed in Chapter 5 must meet this condition. Evolutionary geneticists Joe Felsenstein and John Maynard Smith (Felsenstein, 1988; Maynard Smith, 1988) have proposed that the different theories for the evolution of sex discussed above can all be understood by considering the reasons why they produce nonzero D. Speaking very generally, organisms have two basic kinds of traits, beneficial and deleterious. There are also two basic reasons why traits might be associated with one another: the first is as a result of stochastic forces (in other words, by chance) and the second is as a result of deterministic forces such as natural selection. This leads to a four-way scheme to classify theories of sex based on new combinations of genes. How do the theories discussed in Chapter 5 fit into this four-way scheme?

The Vicar of Bray, in which sex is advantageous because it can combine beneficial traits in the same individual and thereby keep them from competing with one another, is clearly a theory involving beneficial traits, but the theory seems to involve both chance effects and natural selection—which effect is responsible for nonzero D? Is it clear why D must be nonzero for this theory to work? If it weren't zero, that is, if traits were randomly associated, all combinations of traits would exist in their expected proportions. In particular, the combination in which both beneficial traits are present in the same individual would exist and so this combination could be selected.

Consequently, there would be no need for sex if D were zero. Now, what is the cause of nonzero D?

The source of nonrandom associations of traits is chance effects based on finite population size. Recall from the text how we began our consideration of the process by considering the fate of a new beneficial mutation in a population in which other beneficial mutations were also beginning to increase in frequency. We pointed out that it would be very unlikely for both beneficial mutations to exist in the same individual. However, if the population were extremely large, this probability would be realized and some individuals would have both beneficial mutations. Such individuals would be able to increase in frequency, all without the necessity of sex.

Muller's ratchet is clearly a theory involving harmful traits and chance effects, but how is nonzero D created? Although not immediately obvious, nonzero D is again created by chance effects resulting from finite population size. To see this, recall that some mutation classes are absent in a population that has lost a few of its least-loaded mutation classes, that is, a population in which the ratchet has operated. However, if the population were infinite in size, all mutation classes would be present in their randomly expected proportions. The fact that certain kinds of individuals, namely those belonging to the least-loaded mutation class, are less frequent than expected, means that other kinds of individuals, namely those with greater numbers of deleterious mutations, are more common than expected. Put simply, in a finite population, the deleterious traits (or the mutations that cause them) tend to be positively associated with one another.

As it turns out, nonrandom combinations of traits are always generated in finite populations. Since all populations in nature are finite and some quite small, finite population size will always be a contributing factor to nonzero D. Are there other sources of nonzero D, that is, are there sources that would create nonrandom associations of traits even in a large population?

The extension of Muller's ideas on mutation load involving threshold selection clearly involves deleterious genes. Linkage disequilibrium is created by the synergistic nature of selection, whereby genomes with many deleterious genes die and those with few survive. As a consequence good genes tend to be more associated with one another than expected if they were independent.

Natural selection can produce nonzero D in large populations, and the other theories discussed rely on particular forms of natural selection for the production of nonrandom associations of traits. For example, the host-parasite and moving intermediate optimum hypotheses were explicitly expressed in terms of selection favoring alternating combinations of genes at two or more loci, probably because they were formulated after the importance of D was discovered. Selection favoring alternating combinations of genes keeps D away from zero so that sex can continually have an effect. It is less easy to see how the lottery metaphor translates into this terminology.

The lottery concerns beneficial traits and selection, but makes no obvious reference to nonrandom associations between traits. We learned that for the lottery metaphor to work, offspring from the same parent must compete with one another or else they will not be entering the same lottery. If offspring did not interact with one another, they would each be entering their own lottery, and a sexual parent would be like a person who bought 100 different tickets to 100 different lotteries. Offspring from the same parent will tend to have the same associations of genes, and so the requirement that offspring compete with one another translates into the existence of nonrandom associations between genes, that is, nonzero D.

Chapter 6: Twice Nothing

REFERENCES FOR MAIN TEXT

The quotation that begins the chapter is from Benjamin Jowett's translation of Plato's *The Symposium* as it appears in the Library of The Future (Second Edition, Windows™ Version 1.1, Electronically Enhanced Text © Copyright 1991, World Library, Inc.). The idea that sex functions in DNA repair has had a long history. Plato in *The Symposium* was the first to express the general idea in qualitative terms, although according to Freud (1961) the idea is of Babylonian origins. Doughtery (1955) was the first to apply the idea to the evolution of sex. Felsenstein (1974) mentioned the idea in passing; Bernstein (1977) proposed the idea; Maynard Smith discussed the idea briefly in the introduction to his book (1978b); and Walker also considered the general idea (1978). The development given in Chapter 6 comes from the work of my colleagues and I (Bernstein et al., 1989; Bernstein et al., 1987; Bernstein et al., 1985b; Bernstein et al., 1985c; Bernstein et al., 1985a; Bernstein et al., 1984b; Bernstein et al., 1981). The first few paragraphs of Chapter 6 are taken from an article of mine (Michod, 1989). The measurement of DNA damage in humans mentioned in the chapter comes from a discussion and references in Bernstein et al., (1988) and Bernstein et al., (1987). The data showing that recombinational repair functions in DNA repair are summarized in Bernstein (1983), Bernstein et al., (1987) and Bernstein and Bernstein (1991). Meiotic recombination has been studied extensively in fungi and flies (Whitehouse, 1982). Even male Drosophila, which normally have no meiotic recombination (a rarity in nature), undergo both mitotic and meiotic recombination when subjected to DNA damaging agents (Wurgler, 1991; Ferres, et al. 1984; Whittinghill and Lewis, 1961; Ayaki, et al. 1990). The double-strand break repair model was developed to explain this data (Orr-Weaver and Szostak, 1985; Szostak et al., 1983). A nice overview of molecular models of recombination is provided by Devoret (1988). Meiotic recombination and the double-strand break repair model is interpreted from the point of view of the evolution of recombination in Bernstein et al., (1988). Mitotic recombination responds to DNA damage in single-celled organisms like yeast (Kunz and Haynes, 1981) and in multicellular diploid organisms like Drosophila (Martensen and Green, 1976; Wurgler, 1991; Ferres, et al. 1984; Whittinghill and Lewis, 1961; Ayaki, et al. 1990). Mitotic recombination can be detected by the occurrence of twin spots in the skin of Drosophila and segregation patterns in fungi (see any text on genetics, such as Russell [1992]). Twin spots are rare in humans but do occur more frequently in people with certain disorders such as Bloom syndrome (Bernstein and Bernstein 1991). Mitotic recombination in the germ line in Drosophila can be detected by the clustering of recombinants when compared to the expected Poisson distribution (Kidwell and Kidwell, 1975).

An excellent text on microbiology is Davis et al. (1980). An overview of the mechanism of bacterial transformation can be found there or in Wojciechowski (1992). An overview of the evolution of transformation can be found Hudson and Michod (1992). The experiments discussed in the text on *Bacillus subtilis* come from the work in my lab (Hoelzer and Michod, 1991; Michod et al., 1988; Wojciechowski et al., 1989; Michod and Wojciechowski, 1994). Information about the mechanism of transformation *B. subtilis* can be found in many places (see, for example, Anagnostopolous and Spizizen, 1961; deVos and Venema, 1983; Dubnau, 1982; Hadden and Nester, 1968; Nester and Stocker, 1963; Stewart and Carlson, 1986; Yasbin et al., 1975; Spizizen, 1958; Wojciechowski, 1992). Other work on the evolution of transformation in *B. subtilis* and other bacteria include the following references: Cohan et al. (1991); Mongold (1992); Redfield (1993); Redfield (1994); Redfield (1988); Duncan et al., (1989). A computer simulation of the role of deleterious mutation in the evolution of transformation can be found in Redfield (1988). As mentioned in the text, transformation can transfer genes and in so doing, help cells adapt to new environments (Maynard Smith, 1990; Maynard Smith et al., 1991; Graham and Istock, 1979; Graham and Istock, 1978). Multiplicity reactivation is discussed in Bernstein et al. (1987). The work on immunity of infection and DNA damage is from Bernstein (1987).

Chapter 7: Chance and Necessity

REFERENCES FOR MAIN TEXT

The Monod and Democritus quotes are taken from Monod (1971). An estimate of several lethal recessive mutations in the human genome is given in Bittles et al. (1991). The reference for the mutation load and genome size graph is Drake (1974). Gillen and Nossal (1976) show that mutants with lower mutation rate replicate more slowly. A lower mutation rate can also be attained by increased energy use (Hopfield, 1974) or the use of additional gene products (Alberts et al., 1980). Rupp and Howard-Flanders (1968) show that replication past a damage site produces a gap in the complementary strand (see also Bernstein and Bernstein, 1991, 187–189). A discussion of the sickle-cell polymorphism from the population genetics point of view can be found in Templeton (1982). Haldane (1937) first proposed the mutation load principle. The effect of the mode of reproduction on Haldane's mutation load principle was studied by Hopf et al. (1988). The table of diploid modes of reproduction was adapted from Bernstein et al. (1985a). Data indicating a diploid mutation rate of approximately one can be found in Mukai et al. (1972).

Background on Selfing and Outcrossing

Mating system genes are a special kind of sex gene that affects with whom an organism mates. We wish to study two kinds of mating system genes: a gene that promotes selfing and a gene that promotes outcrossing. We assume that the recombination and repair genes are the same in the two mating types. For a sex gene to realize the fruits of its labor, so to speak, it must not only produce a favorable combination of nonsex genes in its offspring (which is a very chancy business), but it, the sex gene, must also be passed on to these same offspring that have the favorable combination nonsex genes. The mating system genes have no direct beneficial effect (the recombinational repair genes are the same in both mating types), and they can only increase by hitching a ride with the favorable gene combinations of nonsex genes that they create. And this makes the whole business of understanding their dynamics very difficult. The relevant technical literature that has studied the dynamics of selfing and outcrossing genes is extensive (Campbell, 1986; Charlesworth et al., 1990; Charlesworth and Charlesworth, 1987; Charlesworth and Charlesworth, 1990; Michod and Gayley, 1994; Michod and Gayley, 1992; Uyenoyama and Waller, 1991a; Uyenoyama and Waller, 1991b; Uyenoyama and Waller, 1991c; Ziehe and Roberds, 1989). In the following discussion I have tried to distill the essential conceptual issues from this work. My presentation here follows closely a section in a recent paper by Todd Gayley and me (Michod and Gayley, 1994).

Inbreeding Depression

As discussed in the text, according to the inbreeding depression rule, we predict selfing to win over outcrossing whenever inbreeding depression is less than one-half. Conversely, we predict outcrossing to win when inbreeding depression of selfing is greater than one-half. Seems simple enough, but, unfortunately, these predictions don't always work: at least that is what recent computer and mathematical models have told us. To understand why, we need to consider the process of natural selection on mating system genes in more detail. First some background on inbreeding depression.

Inbreeding depression can be determined empirically by measuring the fitnesses of two kinds of offspring: those that have been produced by selfing and those produced by outcrossing. In the mathematical models, inbreeding depression is usually calculated from genotype frequencies and fitness parameters, but I think it is easier to think of it as the result of a thought experiment: take all the parents in a population, self them, and measure the fitness of the resulting offspring, then take the same parents, outcross them, and measure the fitness of those offspring. An important point here is that inbreeding depression is an average characteristic of a population.

Now consider the fate of a new type of organism (a new mating system gene, actually) that selfs all the time (never tries sex even once) in a population in which all the other individuals are outcrossing sexually. As we have discussed in the text, both will conduct gene repair just fine. Assume further that mutation is constantly creating deleterious recessive alleles for nonsex genes and thus the original outcrossing parents carry a number of these recessive mutations in the heterozygous (masked) state. We could calculate an inbreeding depression for this population by mathematically performing our thought experiment. Depending on the mutation rate and other parameters chosen, it could easily be much greater than the critical value of one-half, due to unmasking of the recessive deleterious alleles in selfed offspring. However, the mathematical models have shown that even though inbreeding depression is greater than one-half, the selfing individuals will come to dominate the population and outcrossing will disappear. How can this be?

MATING SYSTEM GENES AFFECT THEIR GENETIC BACKGROUND

The new selfer takes a big hit in offspring fitness the first time it reproduces because of its increased homozygosity for fitness determin-

Table 1 **Offspring produced by selfing for a single gene**

parent:		$Aa \times Aa$	
offspring:	$\frac{1}{4}AA$	$\frac{1}{2}Aa$	$\frac{1}{4}aa$
fitness outcome (if the *a* allele is a recessive mutation):	purging	masking	inbreeding depression

ing nonsex genes. This fitness loss is by definition what inbreeding depression measures. However, as a direct result of this hit, the new selfer will purge some deleterious mutations from its genome. Selfed offspring are more homozygous, yes, but more homozygous for *both* alleles at each heterozygous locus—the good dominant allele (*A* in Table 1) and the deleterious recessive allele (*a* in the table). In Table 1, the offspring produced by selfing are listed for just one heterozygous gene, *Aa.* According to the laws of genetics discovered by Gregor Mendel, one in four offspring will be homozygous for the good *A* gene. For these offspring, the bad *a* gene has been purged. If there were two such heterozygous traits, say *AaBb,* encoded by genes on separate chromosomes, the probability of purging both the *a* and *b* mutations would be one in sixteen. If the heterozygous traits are encoded by genes on the same chromosome, the probability of purging depends on the map distance between them. In outcrossing populations there are many such traits that harbor deleterious mutations in heterozygous state. As we know from the text, geneticists have estimated that each of us contains several lethal genes masked in heterozygous state (Bittles et al., 1991). These genes affect critical traits, but the traits affected in each of us are different, so that when we outcross, the lethal genes tend to be kept heterozygous. And these are just the lethal ones: there are many more mutations with less drastic effects hiding in our genomes. The more such genes, the less likely any offspring will purge all the bad genes (be homozygous for the good gene for each trait). Nevertheless, a few offspring may be homozygous for the good gene for at least a few traits and remain heterozygous for the rest. In time with continued selfing all bad genes may be purged.

When the selfer's offspring (the very few that survive by virtue of purging deleterious mutations) grow up to reproduce, the consequences of selfing will be very different from what their parent experienced. Some of the traits that harbored bad genes in heterozygous state will have been purged of the recessive mutations, so the offspring will have fewer heterozygous traits to contend with when they reproduce. This happens every generation, each offspring generation having it better than its parents. Yet the inbreeding depression for the population is unchanged (the single selfer and its descendants are negligible when compared to the whole population), and thus it, inbreeding depression, no longer has any bearing on the future course of evolution of the selfing.

We will refer to the effect that a new mating type has on its own genetic background as its "ripple effect." Inbreeding depression measures fitness consequences only during the first generation of a new mating type's existence, when its genetic background is the same as the rest of the population. Thus, in cases where a new mating type creates a small ripple, we expect to find that the inbreeding depression criterion is close to being a correct predictor of the success of the new kind of mating. Conversely, in cases where the new mating type creates a large ripple around itself, we expect that inbreeding depression will not be a correct predictor of success or failure of the new mating type.

In the case just considered, that of an outcrossing population harboring deleterious recessive mutations, it was easier for selfing to increase than we expected based on considering inbreeding depression and the twofold cost of sex. Are there any cases in which the ripple effect of the new mating type causes outcrossing to fare better than we expect based on the inbreeding depression rule? Yes, when the heterozygote is just plain superior to either homozygote.

We have been discussing heterozygosity for genes for which there are dominant alleles masking recessive mutations. The heterozygote (Aa) performs just like the dominant homozygote (AA). Only the recessive homozygote (aa) does worse. However, for some genes, heterozygotes are superior to *both* homozygotes, by virtue of producing something that neither homozygote can produce. Any homozygosity for these genes is harmful and creates inbreeding depression. Furthermore, purging no longer occurs. If inbreeding depression is caused by genes with superior heterozygotes, outcrossing can increase even though inbreeding depression is less than one-half.

HETEROZYGOTE SUPERIORITY FOR NONSEX GENES HELPS OUTCROSSING

For traits that are caused by heterozygote superiority, as in the cases of the mating type locus in the common baker's yeast (Durand et al. 1993) and malarial resistance in sickle-cell heterozygotes (discussed in the text), computer and mathematical models have shown that more outcrossing may be favored than is currently present in a population even though inbreeding depression is less than the critical value of one-half. In the cases studied, an increase in outcrossing never pays for itself in the generation in which it occurs, because the increased heterozygosity of the offspring cannot make up for the cost of outcrossing. However, by increasing its likelihood of outcrossing, an individual will have injected increased heterozygosity into its lineage of descendants. If the more outcrossed offspring themselves go back to selfing right away, the more active outcrosser's descendants will enjoy these benefits for a while without continuing to pay the costs of sex. The increased heterozygosity among the descendants will gradually diminish to the level of the rest of the population, true, but if there are enough generations of selfing before one of the descendants outcrosses again, it will be possible to more than make up the initial deficit incurred by the more active outcrosser.

Therefore, we expect natural selection to favor individuals that slightly increase their likelihood of outcrossing over that currently present in the population. Indeed, this is what the mathematical and computer models have shown. The degree of increase in outcrossing is crucial to success: the smaller the increase in outcrossing, the better the outcrossing gene's probability of success. Although increased outcrossing is never beneficial in itself, it can be thought of as a strategy for selfers to inject more heterozygosity into their lineage.

A lower initial outcrossing rate also facilitates increasing outcrossing. If the population is predominately selfing, the ripple effect of an increase in outcrossing is larger than if the population already contained significant proportions of both selfers and outcrossers. A predominately selfing population is highly homozygous and an increase in heterozygosity has a larger effect the more homozygous the present genetic background. A lower initial outcrossing rate also facilitates the increase of outcrossing because it reduces the cost of the increase, by raising the inbreeding depression experienced by most of the population (which are predominantly selfers by virtue of their low initial outcrossing rate).

The ripple effect works to the advantage of increased outcrossing (as long as the increase is not too large), so the inbreeding depression criterion is expected to be too stringent in predicting whether more outcrossing will be favored by natural selection. But, as we have found, the magnitude of the increase is critical to the new outcrossing gene's success, small increases being more favorable. To take advantage of the ripple, the new outcrossing gene must refrain from outcrossing again for a while.

What happens to new selfing genes that try to decrease the probability of outcrossing in the population in the case of nonsex genes for which the heterozygote is superior to both homozygotes? What we learned for outcrossing genes in this case applies in reverse for genes that increase the selfing rate. Since we are considering cases in which the inbreeding depression is less than one-half, an act of selfing always has positive fitness consequences. Thus, increased selfing is always advantageous the first time it occurs. However, as a result of an increase in selfing, the offspring will now be more homozygous than the average. If the new type continues to self, each successive generation of selfing will make its background even more homozygous than the rest of the population. The ripple effect is detrimental to the prospects of increased selfing (decreased outcrossing), and the ripple is larger the larger the increase. Therefore, the inbreeding depression criterion is too lenient in predicting the success of increased selfing. Just because you win in the first generation doesn't mean you win in the long run, and small changes in the likelihood of selfing, which take the initial win and then quickly return to the average genetic background, are more likely to invade.

In summary, if fitness is affected by nonsex genes for which the heterozygote is superior to both homozygotes, the ripple effect makes it easier for outcrossing to evolve in small steps, easier than we were led to believe by the inbreeding depression rule discussed in the main text.

RECESSIVE DELETERIOUS MUTATIONS

Let us return to the case of traits in which variation in fitness-determining nonsex genes is maintained by recurrent mutation to deleterious recessive or nearly recessive alleles. Consider first the case of an increase in the selfing rate over what currently exists in the population. We already described the effect that a large change has on its genetic background and how this can lead to more selfing despite a high inbreeding depression. What about small changes in the selfing/outcrossing rate? The advantage of purging your genetic background accrues only to those lineages that stick with selfing long enough to become better off than the rest of the population. As soon as you outcross, you lose your ripple and effectively have to start over again. In other words, new individuals with only a small change in selfing create only a small ripple, so we expect the inbreeding depression criterion to be nearly correct for them. This is what the models have shown.

Now consider new individuals that slightly increase their likelihood of outcrossing over that currently existing in the population. If the mutation rate is not high, then single acts of outcrossing have little effect on their genetic background (provided, of course, that they are not more purged than the rest of the population). Thus, an increase in outcrossing creates little ripple. Gradually, of course, an outcrossing lineage's background will begin to accumulate recessives to a greater degree than will the population, but this occurs on a longer time scale than the selection process because it depends on the mutation rate. If outcrossing is good for you in the first generation, it will be good in subsequent generations, and likewise if it is not good for you. Thus, we expect the inbreeding depression criterion to be nearly correct and expect that this should be true regardless of the size of the change in mating system. This prediction is consistent with computer simulation results from the literature but is in need of more study to be clearly understood.

The premise of using inbreeding depression to predict the evolution of outcrossing and selfing rates is essentially the same as that on which all "strategic" models of evolution are based (for example, arguments based on purely phenotypic considerations such as evolutionary stable strategies [Maynard Smith, 1982]). The basic idea is that we expect type 1 to increase in competition with type 2 when the genotypic fitness of type 1 is greater than the genotypic fitness of type 2. This type of "survival of the fittest" reasoning is valid only when a consistent fitness effect accrues for the two types in all situations.

However, in the case of the evolution of selfing and outcrossing, genes that change the mating system change their own genetic background. So different mating system genotypes experience different fitness consequences for their acts. Thus there is no meaningful "average fitness effect of selfing" that can be applied to every act of selfing, and we expect reasoning based entirely on fitnesses of phenotypes to be incorrect.

This insight makes us more comfortable with two possibilities that have been recently demonstrated for the heterozygote-superiority case by computer simulations and analysis of stability conditions: the existence of selfing rates that resist invasion by any small change, whether it increases or decreases the selfing rate, and also those that allow any kind of change to invade. These results defy "survivel of the fittest" thinking because we know that either outcrossing or selfing should be "better," so why don't modifiers that increase the better strategy increase? The answer is that the increase of a selfing modifier and an outcrossing modifier are not logically complementary processes. Rather, they depend on the specific genetic dynamics induced by the modifier allele, and there is no a priori reason to expect that only one or the other type should ever be able to invade.

OVERVIEW OF SELFING AND OUTCROSSING

The underlying population dynamics of sex and asex genes are complex. Because of this complexity, intuition often fails to predict the outcome of natural selection. Mathematical and computer models, although abstract, have the virtue of making all assumptions explicit and, if the math and computer code is done correctly, of preserving the underlying logic and truth of the conclusions. We encountered this complexity already in Chapter 5, where we considered the hypothesis that sex is beneficial because it creates variability. In the present discussion we have tried to distill from the many recent analyses of mathematical models and computer simulations those conceptual issues that can be made clear. Yet there are other aspects of the models that cannot so easily be put into words. Nevertheless, what have we learned about natural selection and the evolution of sex from these examples?

Natural selection in a sexually reproducing population involves more than just fitness considerations—natural selection is not just "survival of the fittest," as is discussed in Chapter 9. The effect that selfing and outcrossing genes have on genotypic Darwinian fitness can be expressed in terms of the inbreeding depression and cost of sex criterion, yet this fitness criterion misses much of the evolutionary process. We characterized these situations as involving mating system genes that create a large ripple effect on the rest of the genome, that is, on the nonsex genes that are affecting fitness. This ripple effect of mating system genes in effect creates a unique genetic environment in terms of the nonsex genes that are near them on the chromosome. The ripple effect at times helps the outcrossing aspects of sex, most notably for nonsex genes whose effects on fitness result from the heterozygote being superior to both homozygotes. In other cases, in which recessive deleterious mutations are occurring for nonsex genes, it may be more difficult for the outcrossing aspects of sex to increase than purely fitness considerations led us to believe.

Chapter 8: Plato's Theory

REFERENCES FOR MAIN TEXT

The ideas on the function of premeiotic replication are taken from Bernstein et al. (1988) and the references therein. The two-thirds estimate of the frequency of cryptic recombination comes from data in fungi and *Drosophila* (Bernstein et al., 1988; Whitehouse, 1982; Chovnick et al., 1970). Eros Szathmary is the critical colleague and friend who believes mitotic recombination in diploid cells, not sexual recombination, is the solution to the damage problem (Szathmary et al., 1990). His criticism is addressed in two recent papers (Long and Michod, 1994; Michod and Long, 1994). In contrast to the basically cooperative view of sex advanced in this book, evolutionary discussions of sexual behavior often emphasize the antagonistic and aggressive interactions between the sexes (see Ridley [1993] for an overview of the area). The evolution of the haploid-diploid life cycle is discussed in the following references: Adams and Hansche, 1974; Destombe et al., 1993; Kondrashov and Crow, 1991; Lewis, 1985; Long and Michod, 1994; Perrot and Valero, 1991; Michod, 1993; Valero et al., 1993; Michod and Gayley, 1994. The work and quotation involving the structural and DNA sequence homology of recombination proteins is from Story et al. (1993). The idea that chiasmata function in chromosome disjunction comes from Carpenter (1984). Plants protect themselves by producing toxins that damage their predators DNA (Ames, 1983; Bernstein et al., 1988). Outcrossing rates are increased by DNA damage in single-celled organisms (Bernstein and Johns, 1989; Bernstein, 1987; Herskowitz, 1988; Michod and Wojciechowski, 1994). The molecular biologist who has shown that the RecA protein in *E. coli* is designed for DNA repair is the biochemist Michael Cox (Cox, 1993; Cox, 1991).

Chapter 9: Darwinian Dynamics

REFERENCES FOR MAIN TEXT

This chapter contains material first presented in the following papers: Byerly and Michod, 1991a; Byerly and Michod, 1991b; Michod, 1990; Michod, 1981. The *Harper's* magazine article is by Tom Bethell (1976). The propensity definition of fitness is discussed throughout the philosophical literature dealing with fitness (Brandon and Beatty, 1984; Brandon, 1978; Brandon, 1991; Mills and Beatty, 1979; Rosenberg and Williams, 1986; Rosenberg, 1983; Sober, 1984a,b). Henry Byerly and I have criticized the propensity interpretation of fitness for a variety of reasons (Byerly and Michod, 1991a; Byerly and Michod, 1991b), but most importantly, the propensity interpretation doesn't capture the way the term is used in the theory of evolution. Fisher first defined fitness as the per capita rate of increase of a type (Fisher, 1958; Fisher, 1930; see also Bernstein et al., 1984a; Ginzburg, 1983). The definitions of adaptedness given by those quoted in the text are from the following references: Brandon, 1991; Brandon, 1978; Dobzhansky, 1968; Pianka, 1978.

Heterozygote Superiority

Let W_i be the Darwinian fitness of type i, the type's expected number of gametes produced (or offspring in an asexual population). For the moment, we assume that W_i is a true measure of adaptedness of the type. We ask how Fisherian fitness, F_i, depends on W_i, as a way of considering Darwin's phrase "survival of the fittest" in an explicit framework. We consider variation at just a single gene locus with two alleles A and a. This can be any locus we choose. In a diploid organism, like us, there are three genotypes AA, Aa and aa. We refer to these three kinds of organism by the subscript i so that $i = 1,2,3$ stands for genotypes AA, Aa and aa, respectively. In other words, W_2 is the expected number of gametes that an organism with the Aa genotype produces during its lifetime. We census gametes, not offspring, because, in a sexual population, the number of offspring produced depends on the genotypes of the mating pair—the mother and father. The variables and terms defined so far are summarized in Table 2.

We assume that organisms do not choose mates based on their genotype; the organisms are blind with regard to the genotype of their mate. Another way of expressing this is to say that mating occurs at random. We assume random mating, not because it is common in nature (it does occur, sometimes, but organisms often pay a great deal of attention to whom they mate with) but because this makes the mathematics simpler. Finally, we assume that the adaptedness values, the W's, are constant and that the Aa heterozygote is the most adapted.

Table 2

Genotype	AA	Aa	aa
Adaptedness	W_1	W_2	W_3
Fisherian Fitness	F_1	F_2	F_3
Numbers	N_1	N_2	N_3

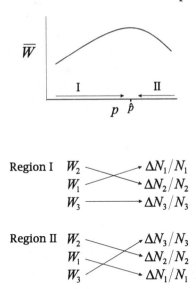

Figure 1. Survival of the fittest is false.
In the top panel, average fitness, $\overline{W}$, is graphed as a function of frequency of the A allele, p, for the case of heterozygote superiority, $W_2 > W_1 > W_3$. Evolution takes the gene frequency to an intermediate value, $\hat{p}$, corresponding to the maximum of average fitness. In region I, $p < \hat{p}$, and gene frequency tends to increase with time. In region II, $p > \hat{p}$, and gene frequency tends to decrease. In the lower panel is shown the relationship between adaptedness and evolutionary success for the two regions of gene frequency. This relationship is discussed further in the text.

We will think of the W's as measures of adaptedness values or "fitness" in Darwin's phrase. The W's are not measures of evolutionary success; they are fixed by assumption, and by assumption the heterozygote is the best, no confusion about that. We consider Fisherian fitness as our operational measure of evolutionary success, "survival" in the phrase "survival of the fittest." Fisherian fitness means the per capita rate of increase, or the change in numbers after a single generation: $\Delta N = N_{t+1} - N_t$. The per capita rate of increase is $\Delta N/N$.

Figure 1 shows the relationship between adaptedness, the W's, and evolutionary success as measured by the per capita rate of increase. The relationship between evolutionary success and adaptedness is different at different stages of the evolutionary process, but at no stage does the heterozygote have the greatest Fisherian fitness, even though it is most adapted. Survival of the fittest predicts the same ranking of evolutionary success and adaptedness. If $W_2 > W_1 > W_3$, then we should have $\Delta N_2/N_2 > \Delta N_1/N_1 > \Delta N_3/N_3$, according to survival of the fittest reasoning. However, this never happens.

When the heterozygote is the most adapted, there is an intermediate gene frequency to which the state of the population tends with time. We need only consider one of the alleles, say the A allele, and let $\hat{p}$ be the intermediate frequency to which evolution tends ($1 - \hat{p}$ is the frequency of the a allele). When the frequency of A is below $\hat{p}$, it increases, and when it is above $\hat{p}$, it decreases. The ranking of evolutionary success of the different genotypes is different in the two regions of gene frequency (below and above $\hat{p}$) as indicated in Figure 1. For each region, the relationship between adaptedness and evolutionary success is given. In both regions the heterozygote is most adapted by assumption, so the ranking of adaptedness must be the same: $W_2 > W_1 > W_3$. The ranking of evolutionary success is different in the two regions, but the heterozygote never enjoys the most success. In region I, the AA genotype has the highest per capita rate of increase, while in region II the aa genotype has the highest per capita rate of increase. In neither region does the heterozygote do the best over time.

PARADIGMS FOR NATURAL SELECTION

We now consider just two types, $i = 1, 2$. We include *intra*specific density effects in the Darwinian dynamic given in the text by multiplying the birth and death probabilities for a type by the type's density raised to the u and v powers, respectively. This yields the following pair of equations

$$\frac{\Delta N_1}{N_1} = b_1 R N_1^u - d_1 N_1^v$$

$$\frac{\Delta N_2}{N_2} = b_2 R N_2^u - d_2 N_2^v$$

$$R = K - N_1 - N_2$$

These equations are identical to the equation given in the text

(p. 160), except for the within-type density-dependent effects on the right-hand side. The effects of density on competitive interactions *between* types is still represented by available resources, R. However, there are now *within* type density-dependent effects, as represented by N^u_i and N^v_i in the equations. What kind of situations might these terms represent?

The case $v < 0$ describes a decrease in the mortality rate as numbers increase. This may occur whenever social interactions are important to survival. For example, defense against predators may be more effective when there are many individuals in the population but less effective as numbers decline. The case $v > 0$ describes an increase in mortality as numbers increase. This might be expected whenever type-specific disease is important: for example, parasites (or predators) may not seek out a host until the numbers of the host get large. The case $u > 0$ describes a decrease in reproduction when rare. This is termed a "cost of rarity." A large number of situations could give rise to a cost of rarity, such as a difficulty of encountering a suitable mate or inbreeding depression. Most mechanisms expected to produce a cost of rarity, $u > 0$, are either directly or indirectly related to sexual reproduction. The consequences of the cost of rarity are profound, especially for the organization of variation in nature and this topic is considered further in Chapter 10. The case $u < 0$ corresponds to factors that may increase reproduction as numbers get rare, factors which are not included in resource competition, R. For example, it has been observed in some fruit fly populations that as a type of male becomes rare, its success at mating increases (Anderson, 1969).

If we now ask when Type 2 can invade a population dominated by Type 1, the answer is similar to the simpler case considered in the text, but there are additional terms involving equilibrium numbers:

$$\hat{N}_2^{u-v}\frac{b_2}{d_2} > \hat{N}_1^{u-v}\frac{b_1}{d_1}.$$

The "hat" above the numbers of the two types refers to the numbers of organisms existing when the new type, type 2, is introduced. Since type 2 did not exist before it was introduced by mutation or migration, $\hat{N}_2 = 0$. $\hat{N}_1$ is the number attained if type 1 were alone and increased to its equilibrium level allowed by resources. We do not need an expression for that number for our purposes.

If $u = v$, we have the condition obtained in the text in which adaptedness, as measured by b_i/d_i predicts evolutionary success or increase of the type. However, the adaptationist position is precarious, since it is true only in the special case of strict equality of u and v which is difficult to satisfy.

If, however, $u > v$, the condition for increase of the new type cannot be satisfied, since the left-hand side of the inequality is zero and the right-hand side is positive. The common type is always stable. By analogy with "survival of the fittest," we describe this situation as "survival of the common," or as "survival of the first." We have stasis. This case stems from sex and is discussed in the text in Chapters 9 and 10. Note that for a sexual species we expect $u = 1$, even when there are no density-dependent effects on survival, $v = 0$.

Finally, consider $u < v$. In this case the inequality is always satisfied, since the left-hand side is infinity. This is the case of "survival of anybody" discussed in the text.

Chapter 10: Darwin's Dilemmas

REFERENCES FOR MAIN TEXT

This chapter presents arguments first given in Bernstein et al. (1985d) and Hopf and Hopf (1985). The study referred to here (and in Chapter 1) as showing an increase in energy allocation to reproduction in plants as density declines is Kodric-Brown et al. (1984). Other examples of the cost of rarity in sexual species can be found in Allee et al. (1949), Andrewartha and Birch (1954), Bernstein et al. (1985d) and Gerritson (1980). In facultatively sexual organisms, sex is usually associated with high density (Bell, 1982; Bernstein et al., 1984a), because the cost of finding a mate is reduced when there are many mates around. References and discussion of the effect of rarity on the cost of mating can be found in Bernstein et al. (1984a).

The distinctness of sexual species has been often debated. In some groups of organisms, especially in plants, workers have claimed that asexual species are as distinct as sexual species (Holman, 1987; Mis-

hler, 1990). Levin (1979), while rejecting the interbreeding definition of species of Mayr (1970), proposes that plant species be
considered as "clusters of diversity in multidimensional character
space." Maynard Smith concluded that "in plants the data on parthenogenetic taxa support the view that discontinuities between species
exist only if there is sexual reproduction" (1983). Sympatric polytypic
clones are commonly found in asexual organisms (Parker, 1979; Vrijenhoek, 1979). An often-cited counter example is the *Bdelloids*, a group
of rotifers in which no male has ever been observed (see, for example,
Hutchinson [1968]). However, after considering the taxonomic literature
on *Bdelloids*, Maynard Smith concluded, "Thus it seems that *Bdelloids*
do not offer any serious challenge to the view that discontinuities between species exist only if there is sexual reproduction" (1983). In spite
of the differing points of view, there can be no doubt about the general
rule: sexual species are far more distinct than asexual taxa (Hutchinson,
1968; Maynard Smith, 1983). The references mentioned for size distri-
butions and character displacement are Elton (1946), Hutchinson
(1959), and Dayan et al. (1989, 1990).

The figures presented in the chapter have been adapted from Bernstein et al. (1985d) and Hopf and Hopf (1985). The work of Mayr on
species is summarized in his books (Mayr, 1970; Mayr, 1982; Mayr,
1963). The data on asexual clones mentioned are from Parker (1979).
The studies of the parthenogenetic weevil are from Suomalainen and
his colleagues (Suomalainen et al., 1976; Suomalainen, 1961). Concerning this extensively studied parthenogenetic weevil, Saura et al. (1976)
summarizes: "It has been shown that an apomictic lineage may diverge
through mutations to a wide array of genotypes, each with slightly
different ecological requirements. . . . Most of the types are rare, occurring in single localities only. Some are again widespread and, interest-
ingly, they seem to inhabit ecologically uniform areas over a
geographically discontinuous area." This is exactly what we expect
based on the top panel of Figure 2 given in the main text. The
description of the mosquito genus *Anopheles* is from Mayr (1970).

Bibliography

Adams, J., and Hansche, P.E. (1974). Population Studies in Microorganisms I. Evolution of Diploidy in *Saccharomyces cerevisiae*. *Genetics* 76, 327–338.

Alberts, B.M., Barry, J., Bedinger, P., Burke, R.L., Hibner, U., Liu, C.-C., and Sheridan, R. (1980). Studies of replication mechanisms with the T4 bacteriophage in vitro system. *In:* Alberts, B., Fox, C.F., and Stusser, F.J. (eds.), *Mechanistic Studies of DNA Replication and Genetic Recombination,* ICN-UCLA Symposia on Molecular and Cellular Biology. Academic Press, San Diego, pp. 449–474.

Allee, W.C., Emerson, A.E., Park, O., and Park, T. (1949). *Principles of Animal Ecology.* Saunders, Philadelphia.

Ames, B.N. (1983). Dietary carcinogens and anticarcinogens. *Science* 221, 1256–1264.

Anagnostopolous, C., and Spizizen, J. (1961). Requirements for transformation in *Bacillus subtilis. J. Bacteriol.* 81, 741–746.

Andersson, M. (1982). Sexual selection, natural selection and quality advertisement. *Biol. J. Linn. Soc.* 17, 375–393.

Andersson, M. (1986). Evolution of condition-dependent sex ornaments and mating preferences: sexual selection based on viability differences. *Evolution* 40, 804–816.

Anderson, W.W. (1969). Polymorphism resulting from the mating advantage of rare male genotypes. *Proc. Nat. Acad. Sci. USA* 64:190–197.

Andrewartha, H.G., and Birch, L.C. (1954). *The Distribution and Abundance of Animals,* University of Chicago Press, Chicago.

Arnold, S.J. (1985). Quantitative genetic models of sexual selection. *Experientia* 41, 1296–1310.

214

Attwood, R.C. (1980). The limits of genetic diversity. *In:* Halvorson, H.O., and Van Holde, K.E. (eds.), *The Origins of Life and Evolution,* New York: Liss, pp. 77–86.

Axelrod, R. (1984). *Evolution of Cooperation.* Basic Books, New York.

Axelrod, R. and Hamilton, W.D. (1981). The evolution of cooperation. *Science* 211, 1290–1296.

Ayaki, T., Fujikawa, K., Ryo, H., Itoh, T., and Kondo, S. (1990). Induced rates of mitotic crossing over and possible mitotic gene conversion per wing anlange cell in *Drosophila melanogaster* by X rays and fission neutrons. *Genetics* 126:157–166.

Bell, G. (1978). The handicap principle in sexual selection. *Evolution* 32, 872–885.

Bell, G. (1982). *The Masterpiece of Nature: The Evolution and Genetics of Sexuality,* University of California Press, Berkeley.

Bell, G. (1988). *Sex and Death in Protozoa.* Cambridge University Press, Cambridge, England.

Bergman, A., and Feldman, M. (1990). More on selection for and against recombination. *Theor. Popul. Biol.* 38, 68–92.

Bernal, J.D. (1967). *The Origins of Life.* World Publishing, Cleveland.

Bernstein, C. (1979). Why are babies young? Meiosis may prevent aging of the germ line. *Perspect. Biol. Med.* 22, 539–544.

Bernstein, C. (1987). Damage in DNA of an infecting phage T4 shifts reproduction from asexual to sexual allowing rescue of its genes. *Genet. Res.* 49, 183–189.

Bernstein, C., and Bernstein, H. (1991). *Aging, Sex and DNA Repair.* Academic Press, San Diego.

Bernstein, C., and Johns, V. (1989). Sexual Reproduction as a Response to Damage in *Schizosaccharomyces pombe. J. Bact.* 171, 1893–1897.

Bernstein, H. (1977). Germ line recombination may be primarily a manifestation of DNA repair processes. *J. Theor. Biol.* 69, 371–380.

Bernstein, H., Byers, G.S., and Michod, R.E. (1981). Evolution of sexual reproduction: importance of DNA repair, complementation and variation. *Am. Nat.* 117, 537–549.

Bernstein, H. (1983). Recombinational repair may be an important function of sexual reproduction. *BioScience* 33, 326–331.

Bernstein, H., Byerly, H., Hopf, F., and Michael, R.E. (1983). The Darwinian dynamic. *Quart. Rev. Biol.* 58:185–207.

Bernstein, H., Byerly, H., Hopf, F., and Michod, R.E. (1984a). Sex and the emergence of species. *J. Theor. Biol.* 117, 665–690.

Bernstein, H., Byerly, H.C., Hopf, F.A., and Michod, R.E. (1984b). Origin of sex. *J. Theor. Biol.* 110, 323–351.

Bernstein, H., Byerly, H., Hopf, F., and Michod, R.E. (1985a). The evolutionary role of recombinational repair and sex. *Int. Rev. Cytol.* 96, 1–28.

Bernstein, H., Byerly, H., Hopf, F., and Michod, R.E. (1985b). DNA damage, mutation and the evolution of sex. *Science* 229, 1277–1281.

Bernstein, H., Byerly, H.C., Hopf, F.A., and Michod, R.E. (1985c). Sex and the emergence of species. *J. Theor. Biol.* 117, 665–690.

Bernstein, H., Byerly, H.C., Hopf, F.A., and Michod, R.E. (1985d). DNA repair and complementation: The major factors in the origin and maintenance of sex. In Halvorsen, H.O. (ed.), *In Origin and Evolution of Sex.* Liss, New York, pp. 29–45.

Bernstein, H., Hopf, F.A., and Michod, R.E. (1987). The molecular basis of the evolution of sex. *Adv. Genet.* 24, 323–370.

Bernstein, H., Hopf, F.A., and Michod, R.E. (1988). Is meiotic recombination an adaptation for repairing DNA, producing genetic variation, or both? *In:* Michod, R.E., and Levin, B. (eds.), *The Evolution of Sex: An Examination of Current Ideas.* Sinauer, Sunderland, MA, pp. 139–160.

Bernstein, H., Hopf, F., and Michod, R.E. (1989). The evolution of sex: DNA repair hypothesis. *In:* Rasa, A., Vogel, C., and Voland, E. (eds.), *Sociobiology of Reproduction, Strategies in Animal and Man.* Bechenham, England, Croom Helm LTD, pp. 3–18.

Bethell, T. (1976). Darwin's Mistake. Harper's, February, 70–75.

Bittles, A.H., Mason, W.M., Greene, J., and Appaji Rao, N. (1991). Reproductive behavior and health in consanguineous marriages. *Science* 252, 789–793.

Bowler, P. (1984). *Evolution: The History of an Idea.* University of California Press, Berkeley.

Boyd, R. and Lorberbaum, J. (1987). No pure strategy is evolutionarily stable in the repeated Prisoner's Dilemma game. *Nature* 327, 58–59.

Brandon, R. (1978). Adaptation and evolutionary theory. *Studies in the History and Philosophy of Sciences* 9, 188–206.

Brandon, R.N. (1991). *Adaptation and Environment.* Princeton University Press, Princeton.

Brandon, R., and Beatty, J. (1984). The propensity interpretation of fitness—no interpretation is no substitute. *Phil. of Sci.* 51, 342–347.

Brown, J.S., Sanderson, M.J., and Michod, R.E. (1982). Evolution of social behavior by reciprocation. *J. Theor. Biol.* 99, 319–339.

Bulfinch, T. (1971) *Bulfinch's Mythology.* HarperCollins, New York.

Byerly, H.C., and Michod, R.E. (1991a). Fitness and evolutionary explanation: a response. *Bio. and Phil.* 6, 45–53.

Byerly, H.C., and Michod, R.E. (1991b). Fitness and evolutionary explanation. *Biol. and Phil.* 6, 1–22.

Cairns-Smith, A.G. (1982). *Genetic Takeover and the Mineral Origins of Life,* Cambridge University Press, Cambridge.

Campbell, R.B. (1986). The interdependence of mating structure and inbreeding depression. *Theor. Popul. Biol.* 30, 232–244.

Carpenter, A.T.C. (1984). Meiotic roles of crossing-over and of gene conversion. *Cold Spring Harbor Symp. Quant. Biol.* 49, 23–29.

Charlesworth, B. (1976). Recombination modification in a fluctuating environment. *Genetics* 83, 181–195.

Charlesworth, D., Morgan, M., and Charlesworth, B. (1990). Inbreeding depression, genetic load, and the evolution of outcrossing rates in a multilocus system with no linkage. *Evolution* 44, 1469–1489.

Charlesworth, D. and Charlesworth, B. (1987). Inbreeding depression and its evolutionary consequences. *Annu. Rev. Ecol. Syst.* 18, 237–268.

Charlesworth, D. and Charlesworth, B. (1990). Inbreeding depression with heterozygote advantage and its effect on selection for modifiers changing the outcrossing rate. *Evolution* 44, 870–888.

Chovnick, A., Ballantyne, G.H., Baillie, D.L., and Holm, D.G. (1970). Gene conversion in higher organisms: half-tetrad analysis of recombination within the rosy cistron of *Drosophila melanogaster. Genetics* 66, 315–329.

Cohan, R.M., Roberts, M.S., and King, E.C. (1991). The potential for genetic exchange by transformation within a natural population of *Bacillus subtilis. Evolution* 45, 1393–1421.

Cox, M.M. (1991). The RecA protein as a recombinational repair system. *Mol. Microbiol.* 5, 1295–1299.

Cox, M.M. (1993). Relating biochemistry to biology: how the recombinational repair function of the RecA protein is manifested in its molecular properties. *BioEssays* 15, 617–623.

Crow, J.F., and Kimura, M. (1970). An Introduction to Population Genetics Theory. Burgess, Minneapolis.

Darwin, C. (1859). *The Origin of Species.* John Murray, London.

Darwin, C. (1889). *The Effects of Cross and Self Fertilisation in the Vegetable Kingdom,* Appleton, New York.

Davis, B.D., Dulbecco, R., Eisen, H.N., and Ginsberg, H.S. (1980). *Microbiology,* 3rd edition. Harper and Row, Philadelphia.

Davis, G.W.F., and O'Donald, P. (1976). Sexual selection for a handicap: a critical analysis of Zahavi's model. *J. Theor. Biol.* 57, 345–354.

Dawkins, R. (1976). *The Selfish Gene.* Oxford University, Oxford.

Dayan, T., Simberloff, D., Tchnernov, E., and Yom-Tov, Y. (1989). Inter- and intra-specific character of displacement in mustelids. *Ecology* 70: 1526–39.

Dayan, T., Simberloff, D., Tchnernov, E., and Yom-Tov, Y. (1990). Feline canines: Community-wide character displacement in the small cats of Israel. *Am. Nat.* 136: 39–60.

Destombe, C., Godin, H., Nocher, M., Richerd, S., and Valero, M. (1993). Differences in response between haploid and diploid isomorphic phases of *Gracilaria verrucosa* (Rhodophyta: Gigartinales) exposed to artificial environmental conditions. *Hydrobioilogia* 260/261, 131–137.

Devoret, R. (1988). Molecular aspects of genetic recombination. *In:* (Michod, R.E. and Levin, B. (eds.), *The Evolution of Sex: An Examination of Current Ideas.* Sinauer Associates, Sunderland, MA, pp. 24–44.

deVos, W.M., and Venema, G. (1983). Transformation of *Bacillus subtilis* competent cells: identification and regulation of the recE gene product. *Mol. Gen. Genet.* 190, 56–64.

Dobzhansky, T. (1968). On some fundamental concepts of Darwinian biology. *In:* Hecht, M.K. and Steere, N.C. (eds.), *Evolutionary Biology.* 2nd edition. Appleton-Century-Crofts, New York.

Dougherty, E.C. (1955). Comparative evolution and the origin of sexuality. *Syst. Zool.* 4, 145–169.

Drake, J.W. (1974). The role of mutation in bacterial evolution. *Symp. Soc. Gen. Microbiol.* 24, 41–58.

Dubnau, D. (1982). Genetic transformation in *Bacillus subtilis. In:* Dubnau, D. (ed.), *The Molecular Biology of the Bacilli,* vol. 1. Academic Press, New York, pp. 148–175.

Duncan, K.E., Istock, C.A., Graham, J.B., and Ferguson, N. (1989). Genetic exchange between *Bacillus subtilis* and *Bacillus cheniformis:* variable hybrid stability and the nature of bacterial species. *Evolution* 43, 1585–1609.

Durand, J., Birdsell, J., and Willis, C. (1993). Pleiotropic effects of heterozygosity at the mating-type locus of the yeast *Saccharomyces cerevisiae* on repair, recombination and transformation. *Mutat. Res.* 290, 239–247.

Eigen, M., Gardiner, W., Schuster, P., and Winkler-Oswatitsch, R. (1981). The origin of genetic information. *Sci. Am.* 244, 88–118.

Eigen, M. (1992). *Steps Towards Life.* Oxford University Press, Oxford.

Eigen, M., and Schuster, P. (1979). The Hypercycle, a Principle of Natural Self-Organization. Springer-Verlag, Berlin.

Elton, C.S. (1946). Competition and the structure of ecological communities. *J. Anim. Ecol.* 15:54–68.

Eshel, I. (1978). On the handicap principle—a critical defense. *J. Theor. Biol.* 70, 245–250.

Felsenstein, J. (1974). The evolutionary advantage of recombination. *Genetics* 78, 737–756.

Felsenstein, J. (1988). Sex and the evolution of recombination. *In:* Michod, R.E. and Levin, B. (eds.), *The Evolution of Sex: An Examination of Current Ideas.* Sinauer, Sunderland.

Ferres, M.D., Alba, P., Xamena, N., Creus, A., and Marcos, R. (1984). Induction of male recombination in *Drosophila melanogaster* by chemical treatment. *Mutat. Res.* 126:245–250.

Ferriere, R., and Michod, R.E. (1994a). Invading wave of cooperation in a spatial iterated Prisoner's Dilemma. *Proc. R. Soc. London* (B) In press.

Ferriere, R., and Michod, R.E. (1994b.) The evolution of cooperation in spatially heterogeneous populations. (submitted manuscript).

Fisher, R.A. (1915). The evolution of sexual preferences. *Eugen. Rev.* 7, 184–192.

Fisher, R.A. (1930). *The Genetical Theory of Natural Selection.* Clarendon, Oxford.

Fisher, R.A. (1958). *The Genetical Theory of Natural Selection.* Dover, New York.

Freud, S. (1961). *Beyond the Pleasure Principle.* Norton, New York.

Gerritson, J. (1980). Sex and parthenogenesis in sparse populations. *Am. Nat.* 115, 718–742.

Gillen, F.D., and Nossal, N.G. (1976). Control of mutation frequency by bacteriophage T4 DNA polymerase I. The tsCB120 antimutator DNA polymerase is defective in strand displacement. *J. Biol. Chem.* 251, 5219–5224.

Ginzburg, L.R. (1983). *Theory of Natural Selection and Population Growth.* Benjamin/Cummings, Menlo Park, CA.

Gould, S.J. and Lewontin, R.C. (1979). The spondrels of San Marco and the Panglossian paradigm: A critique of the adaptationist programme. *In:* Maynard Smith, J. and Holliday, R. (eds.), *The Evolution of Adaptation by Natural Selection.* Royal Society, London. 147–598.

Graham, J.B., and Istock, C.A. (1978). Genetic exchange in *Bacillus subtilis* in soil. *Mol. Gen. Genet.* 166, 287–290.

Graham, J.B., and Istock, C.A. (1979). Gene exchange and natural selection cause *Bacillus subtilis* to evolve in soil culture. *Science* 204, 637–639.

Hadden, C., and Nester, E.W. (1968). Purification of competent cells in the *Bacillus subtilis* transformation system. *J. Bacteriol.* 95, 876–885.

Haldane, J.B.S. (1937). The effect of variation on fitness. *Am. Nat.* 71, 337–349.

Hamilton, W.D. (1963). The evolution of altruistic behavior. *Am. Nat.* 97, 354–356.

Hamilton, W.D. (1964a). The genetical evolution of social behaviour. I. *J. Theor. Biol.* 7, 1–16.

Hamilton, W.D. (1964b). The genetical evolution of social behaviour. II. *J. Theor. Biol.* 7, 17–52.

Hamilton, W.D., Henderson, P.A., and Moran, N.A. (1980). Fluctuations of environment and coevolved antagonist polymorphism as factors in the maintenance of sex. In Alexander, R.D. and Tinkle (ed): *Natural Selection and Social Behavior: Recent Research and New Theory.* Chiron Press, New York. pp. 363–382.

Hamilton, W.D. (1966). The moulding of senescence by natural selection. *J. Theor. Biol.* 12, 12–45.

Hamilton, W.D. (1980). Sex versus non-sex versus parasite. *Oikos* 35, 282–290.

Hamilton, W.D. (1993). Haploid dynamic polymorphism in a host with matching parasites: effects of mutation/subdivision, linkage, and patterns of selection. *J. Heredity* 84, 328–338.

Hamilton, W.D., and Zuk, M. (1982). Heritable true fitness and bright birds: a role for parasites? *Science* 218, 384–387.

Hardin, G. (1968). The tragedy of the commons. Science 162, 1243–1248.

Hasson, O. (1989). Amplifiers and the handicap principle in sexual selection: a different emphasis. *Proc. R. Soc. Lond. {B}* 235, 383–406.

Hedrick, P.W. (1983). *Genetics of Populations.* Science Books, Boston.

Hedrick, P.W., Jain, S., and Holden, L. (1978) Multilocus systems in evolution. *Evolutionary Biology* 11, 101–182.

Hendry, L.B. Bransome, E.D., Jr., Hutson, M.S., and Campbell, L.K. (1981a). First approximation of a stereochemical rationale for the genetic code based on the topography and physicochemical properties of "cavities" constructed from models of DNA. *Proc. Natl. Acad. Sci. USA* 78, 7440–7444.

Hendry, L.B. Bransome, E.D., Jr., and Petersheim, M. (1981b). Are there structural analogies between amino acids and nucleic acids? *Origins of Life* 11, 203–221.

Hesiod and Theognis. (1973). *Hesiod and Theognis.* Penguin, London.

Herskowitz, I. (1988). The life cycle of the yeast *Saccharomyces cerevisae. Microbiol. Rev.*

Hickey, D.A. (1982). Selfish DNA: A sexually transmitted nuclear parasite. *Genetics* 101, 519–531.

Hickey, D.A. (1993). Molecular symbionts and the evolution of sex. *Heredity* 84, 410–414.

Hickey, D.A. and Rose, M.R. (1988). *In:* Michod, R.E. and Levin, B. (eds.), *The Evolution of Sex: An Examination of Current Ideas.* Sinauer, Sunderland, MA, pp. 161–175.

Hoekstra, R.F. (1987). The evolution of sexes. *In:* Stearns, S.C., (ed.), *The Evolution of Sex and Its Consequences.* Birkhauser Verlag, Basel, pp. 59–91.

Hoelzer, M.A., and Michod, R.E. (1991). DNA repair and the evolution of transformation in *Bacillus subtilis.* III. Sex with damage DNA. *Genetics* 128, 215–223.

Holman, E.W. (1987). Recognizability of sexual and asexual species of rotifers. *Syst. Zool.* 36, 381–386.

Hopf, F., Michod, R.E., and Sanderson, M.J. (1988). The effect of reproductive system on mutation load. *Theor. Popul. Biol.* 33, 243–265.

Hopf, F.A. and Hopf, F.W. (1985). The role of the Allee effect in species packing. *Theor. Popul. Biol.* 27, 27–50.

Hopfield, J.J. (1974) Kinetic proofreading: a new mechanism for reducing errors in biosynthetic processes requiring high specificity. *Proc. Nat. Acad. Sci. USA* 71, 4135–4139.

Hudson, R., and Michod, R.E. (1992). DNA transformation, evolution. *In:* Lederberg, J. (ed.), *The Encyclopedia of Microbiology.* Academic Press, San Diego.

Hutchinson, G.E. (1959). Homage to Santa Rosalia or why are there so many kinds of animals. *Am. Nat.* 93: 145–159.

Hutchinson, G.E. (1968) When are species necessary? *In:* Lewontin, R.C. (ed.), *Population Biology and Evolution.* Syracuse University Press, Syracuse, NY, 177–186.

Kidwell, M.G. and Kidwell, J.F. (1975). Spontaneous male recombination and mutation in isogenic-derived chromosomes of *Drosophila melanogaster. J. Hered.* 66:367–375.

Kirkpatrick, M. (1982). Sexual selection and the evolution of female choice. *Evolution* 36, 1–12.

Kirkpatrick, M. (1986a). Sexual selection and cycling parasites: a simulation study of Hamilton's hypothesis. *J. Theor. Biol.* 119, 263–271.

Kirkpatrick, M. (1986b). The handicap mechanism of sexual selection does not work. *Am. Nat.* 127, 222–240.

Kodric Brown, A., and Brown, J.H. (1984). Truth in advertising: the kind of traits favored by sexual selection. *Am. Nat.* 124, 309–323.

Kodric-Brown, A., Brown, J.H. Byers, G.S., and Gori, D.F. (1984). Organization of a tropical island community of hummingbirds and flowers. *Ecology* 65, 1358.

Kondrashov, A.S. (1982). Selection against harmful mutations in large sexual and asexual populations. *Genet. Res.* 40, 325–332.

Kondrashov, A.S. (1984). Deleterious mutations as an evolutionary factor. I. The advantage of recombination. *Genet. Res. Camb.* 44, 199–247.

Kondrashov, A.S. (1985). Deleterious mutations as an evolutionary factor. II. Facultative apomixis and selfing. *Genetics* 111, 635–653.

Kondrashov, A.S. (1988). Deleterious mutations and the evolution of sexual reproduction. *Nature* 336, 435–440.

Kondrashov, A.S., and Crow, J.F. (1991). Haploidy or diploidy: which is better? *Nature* 351, 313–314.

Kunz, B.A., and Haynes, R.H. (1981). Phenomenology and genetic control of mitotic recombination in yeast. *Annu. Rev. Genet.* 15, 57–89.

Lack, D.L. (1947). *Darwin's Finches.* Cambridge University Press, Cambridge.

Lande, R. (1981). Models of speciation by sexual selection on polygenic traits. *Proc. Nat. Acad. Sci. USA* 78, 3721–3725.

Langley, C.H. (1977). Nonrandom associations between allozymes in natural populations of *Drosophila melanogaster. In:* Christiansen, F.B. and Fenchel, T.M. (eds.), *Measuring Selection in Natural Populations,* Springer-Verlag, Berlin, pp. 265–273.

Langley, C.H., Tobari, Y.N., and Kojima, K. (1974). Linkage disequilibrium in natural populations of *Drosophila melanogaster. Genetics* 78, 921–936.

Levin, D.A. (1979). The nature of plant species. *Science* 204, 381–384.

Lewis, W.M. (1985). Nutrient scarcity as an evolutionary cause of haploidy. *Am. Nat.* 125, 692–701.

Lewis, W.M. (1987). The cost of sex. *In:* Stearns, S.C. (ed.), *The Evolution of Sex and Its Consequences.* Birkhauser Verlag, Basel, pp. 33–57.

Lohrmann, R., Bridson, P.K., and Orgel, L.E. (1980). Efficient metal-ion catalyzed template-directed oligonucliotide synthesis. *Science* 208, 1464–1465.

Lohrmann, R., and Orgel, L.E. (1980). Efficient catalysis of polycytidylic acid-directed oligoguanylate formation by PB^{2+}. *J. Mol. Biol.* 142, 555–567.

Lombardo, M.P. (1985). Mutual restraint in tree swallows: a test of the TIT FOR TAT model of reciprocity. *Science* 227, 1363–1365.

Long, A., and Michod, R.E. (1994). Origin of sex for error repair. I. Sex versus diploidy versus haploidy? *Theor. Popul. Biol.* (in press)

Marchant, J. (1975). *Alfred Russell Wallace Letters, and Reminiscences.* Arno Press, New York.

Martensen, D.V. and Green, M.M. (1976). UV-induced mitotic recombination in somatic cells of *Drosophila melanogaster. Mutat. Res.* 36, 391–396.

Maynard Smith, J. (1959). Sex-limited inheritance of longevity in *Drosophila subobscura. J. Genet.* 56, 1–9.

Maynard Smith, J. (1964). Group selection and kin selection. *Nature* 201, 145–147.

Maynard Smith, J. (1976a). A short-term advantage for sex and recombination through sib-competition. *J. Theoret. Biol.* 63, 245–258.

Maynard Smith, J. (1976b). Sexual selection and the handicap principle. *J. Theor. Biol.* 57, 239–242.

Maynard Smith, J. (1977). The sex habit in plants and animals. *In:* Christiansen, F.B. and Fenchel, T.M. (eds.), *Measuring Selection in Natural Populations.* Springer-Verlag, Berlin, pp. 265–273.

Maynard Smith, J. (1978a). The handicap principle—a comment. *J. Theor. Biol.* 70, 251–252.

Maynard Smith, J. (1978b). *The Evolution of Sex.* Cambridge University Press, Cambridge.

Maynard Smith, J. (1980). Selection for recombination in a polygenic model. *Genet. Res., Camb.* 35, 269–277.

Maynard Smith, J. (1982). *Evolution and the Theory of Games.* Cambridge University Press, London.

Maynard Smith, J. (1983). The genetics of stasis and punctuation. *Annu. Rev. Genet.* 17, 11–25.

Maynard Smith, J. (1985a). Sexual selection, handicaps and true fitness. *J. Theor. Biol.* 115, 1–8.

Maynard Smith, J. (1985b). Mini review. Sexual selection, handicaps and true fitness. *J. Theor. Biol.* 115, 1–8.

Maynard Smith, J. (1988). The evolution of recombination. *In:* Michod, R.E. and Levin, B.R. (eds.), *The Evolution of Sex, An Examination of Current Ideas.* Sinauer, Sunderland, MA, pp. 106–125.

Maynard Smith, J. (1990). The evolution of prokaryotes: does sex matter? *Annu. Rev. Ecol. Syst.* 21, 1–12.

Maynard Smith, J., Dowson, C.G., and Spratt, B.G. (1991). Localized sex in bacteria. *Nature* 249, 29–31.

Mayr, E. (1963). *Animal Species and Evolution*. Belknap Press of Harvard University Press, Cambridge, MA.

Mayr, E. (1970). *Populations, Species, and Evolution*. Belknap Press of Harvard University Press, Cambridge, MA.

Mayr, E. (1982). *The Growth of Biological Thought: Diversity, Evolution, and Inheritance*. Harvard University Press, Cambridge, MA.

Michod, R.E. (1981). Positive heuristics in evolutionary biology. *Brit. J. Philos. Sci.* 32, 1–36.

Michod, R.E. (1982). The theory of kin selection. *Ann. Rev. Ecol. Syst.* 13, 23–55.

Michod, R.E. (1983). Population biology of the first replicators: on the origin of the genotype, phenotype and organism. *Am. Zool.* 23, 5–14.

Michod, R.E., and Sanderson, M.J. (1985). Behavioural structure and the evolution of social behaviour. In: Greenwood. J.J. and Slatkin, M. (eds), *Evolution—Essays in Honour of John Maynard Smith*. Cambridge University Press, Cambridge, pp. 95–104.

Michod, R.E., Wojciechowski, M.F., and Hoelzer, M.A. (1988). DNA repair and the evolution of transformation in the bacterium *Bacillus subtilis*. *Genetics* 118, 31–39.

Michod, R.E., Wojciechowski, M.F., and Hoelzer, M.A. (1989). Evolution of sex in prokaryotes. *In:* Clegg, M., and Clark, S. (eds.) *Molecular Evolution*, UCLA Symposia on Molecular and Cellular Biology, New Series, vol. 122.

Michod, R.E. (1989). What's love got to do with it? The solution to one of evolution's greatest riddles. *The Sciences*, May/June, 22–27.

Michod, R.E. (1990). Sex and evolution. *In:* Nadel, L., and Stein, D. (eds.), *1990 Lectures in Complex Systems*, SFI Studies in the Sciences of Complexity, Lect. Vol. III. Addison-Wesley, Reading, MA, pp. 285–320.

Michod, R.E. (1993). Genetic error, sex and diploidy, *J. Heredity* 84, 360–371.

Michod, R.E., and Gayley, T.W. (1992). Masking of mutations and the evolution of sex. *Am. Nat.* 139, 706–734.

Michod, R.E., and Gayley, T.W. (1994). Genetic error, heterozygosity and the evolution of the sexual cycle. *In:* Kirkpatrick, M. (ed.), *Some Mathematical Questions in Biology: Evolution of Haploid and Diploid Life Cycles*. American Mathematical Society.

Michod, R.E. and Hasson, O. (1990). On the evolution of reliable indicators of fitness. *Am. Nat.* 135, 788–808.

Michod, R.E., and B.R. Levin (eds.) (1988). *The Evolution of Sex, an Examination of Current Ideas.* Sinauer, Sunderland, MA.

Michod, R.E., and Long, A. (1994). Origin of sex for error repair. II. Can sex cope with rarity or extreme environments? *Theor. Popul. Biol.*

Michod, R.E., and Wojciechowski, M.F. (1994). DNA repair and the evolution of transformation. IV. DNA damage increases transformation. *J. Evol. Biol.* 7, 147–175.

Miller, S.L., and Orgel, L.E. (1974). *The origins of life on earth.* Prentice-Hall, Englewood Cliffs, NJ.

Mills, S., and Beatty, J. (1979). The propensity interpretation of fitness. *Phil. Sci.* 46, 263–288.

Mishler, B.D. (1990). Reproductive biology and species distinctions in the moss genus *Tortula,* as represented in Mexico. *Syst. Bot.* 15, 86–97.

Mongold, J.A. (1992). DNA repair and the evolution of transformation in *Haemophilus influenzae. Genetics* 132, 893–898.

Monod, J. (1971). *Chance and Necessity: An Essay on the Natural Philosophy of Modern Biology.* Vintage-Random House, New York.

Morowitz, H.J. (1970). *Entropy for Biologists: An Introduction to Thermodynamics.* Academic Press, New York.

Mukai, T., Chigusa, S.I., Mettler, L.E., and Crow, J.F. (1972). Mutation rate and dominance of genes affecting viability in *Drosophila melanogaster. Genetics* 72, 335–355.

Muller, H.J. (1932). Some genetic aspects of sex. *Am. Nat.* 66, 118–138.

Muller, H.J. (1964). The relation of recombination to mutational advance. *Mutat. Res.* 1, 2–9.

Nester, E.W., and Stocker, B.A.D. (1963). Biosynthetic latency in early stages of deoxyribonucleic acid transformation in *Bacillus subtilis. J. Bacteriol.* 86, 785–796.

Nowak, M.A. and Sigmund, K. (1992). Tit for tat in heterogeneous populations. *Nature* 355, 250–253.

Nunney, L. (1989). The maintenance of sex by group selection. *Evolution* 43, 245–257.

Nur, N., and Hasson, O. (1984). Phenotypic plasticity and the handicap principle. *J. Theor. Biol.* 110, 275–297.

Oparin, A.I. (1938). *Origin of Life.* Dover, New York.

Oparin, A.I. (1965). The pathways of the primary development of metabolism and artificial modeling in coacervate drops. *In:* Fox, S.W. (ed.), *The Origins of Prebiological Systems.* Academic Press, New York, pp. 331–346.

Orr-Weaver, T.L., and Szostak, J.W. (1985). Fungal recombination. *Microbiol. Rev.* 49, 33–58.

Parker, E.D. (1979). Ecological implications of clonal diversity in parthenogenetic morphospecies. *Am. Zool.* 19, 753–762.

Perrot, V. Richerd, S., and Valero, M. (1991). Transition from haploidy to diploidy. *Nature* 351, 315–317.

Pianka, E.R. (1978). *Evolutionary Ecology,* 2nd edition. Harper & Row, New York.

Plato (1993) *The Symposium.* Translated by Walter Hamilton. Viking Penguin, New York.

Pomiankowski, A. (1987a). The costs of choice in sexual selection. *J. Theor. Biol.* 128, 195–218.

Pomiankowski, A. (1987b). Sexual selection: the handicap principle does work—sometimes. *Proc. R. Soc. Lond. {B}* 231, 145.

Redfield, R. (1993). Evolution of natural transformation: testing the DNA repair hypothesis in *Bacillus subtilis* and *Haemophilus influenzae. Genetics* 133, 755–761.

Redfield, R. (1994). Genes for breakfast: the have-your-cake-and-eat-it—too of bacterial transformation. *J. Heredity* 84, 400–404.

Redfield, R.J. (1988). Evolution of bacterial transformation: is sex with dead cells ever better than no sex at all? *Genetics* 119, 213–221.

Ridley, M. (1993). *The Red Queen, Sex and the Evolution of Human Nature.* Macmillan, New York.

Rose, M.R. (1983). The contagion mechanism for the origin of sex. *J. Theor. Biol.* 101, 137–146.

Rosenberg, A. (1983). Fitness. *J. Philosophy* 80, 457–473.

Rosenberg, A., and Williams, M. (1986). Fitness as Primitive and Propensity. *Phil. Sci.* 53, 412–418.

Roughgarden, J. (1979). *Theory of Population Genetics and Evolutionary Ecology: An Introduction.* Macmillan, New York.

Rupp., W.D., and Howard-Flanders, P. (1968). Discontinuities in DNA synthesized in an excision defective strain of *Escherichia coli* following ultraviolet irradiation. *J. Mol. Biol.* 31, 291–304.

Russell, P.J. (1992). *Genetics,* 3rd edition. Harper Collins, New York.

Sasaki, A., and Iwaas, Y. (1987). Optimal recombination rate in fluctuating environments. *Genetics* 115, 377–388.

Saura, A., Lokki, J., Lankinen, P., and Suomalainen, E. (1976). Genetic polymorphism and evolution in parthenogenetic animals III. *Otiorrhynchus scaber* (Coleoptera: Curculionidae). *Hereditas* 82, 79–100.

Schuster, P. (1980). Prebiotic evolution. *In:* Gutfreund, H. (ed.), *Biochemical Evolution.* Cambridge University Press, Cambridge, p. 15.

Seger, J., and Hamilton, W. (1988). Parasites and sex. *In:* Michod, R.E. and Levin, B. (eds.), *The Evolution of Sex: An Examination of Current Ideas.* Sinauer, Sunderland, MA, pp. 176–93.

Sober, E. (1984a). Fact, Fiction and Fitness. *Jour. Phil.* 84, 372–383.

Sober, E. (1984b). *The Nature of Selection.* MIT Press, Cambridge, MA.

Spizizen, J. (1958). Transformation of biochemically deficient strains of *Bacillus subtilis* by deoxyribonucleic acid. *Proc. Natl. Acad. Sci. USA* 44, 1072–1078.

Stanley, S.M. (1976). Clades versus clones in evolution: why we have sex. *Science* 190, 282–283.

Stearns, S.C. (ed) (1987). *The Evolution of Sex and Its Consequences.* Birkhauser Verlag, Basel.

Stearns, S.C. (1992). *The Evolution of Life Histories.* Oxford University Press, Oxford.

Stewart, G.J., and Carlson, C.A. (1986). The biology of natural transformation. *Annu. Rev. Microbiol.* 40, 211–235.

Story, R.M., Bishop, D.K. Kleckner, N., and Steitz, T.A. (1993). Structural relationship of bacterial RecA proteins to recombination proteins from bacteriophage T4 and yeast. *Science* 259, 1892–1896.

Suomalainen, E. (1961). On morphological differences and evolution of different polyploid parthenogenetic weevil populations. *Hereditas* 47, 309–341.

Suomalainen, E., Saura, A., and Lokki, P. (1976). Evolution of parthenogenetic insects. *Evol. Biol.* 9, 209–257.

Szathmary, E., Scheuring, I., Kotsis, M., and Gladkin, I. (1990). Sexuality of eukaryotic unicells: Hyperbolic growth, coexistence of facultative parthenogens, and the repair hypothesis. *In:* Maynard Smith, J. and Vida, G. (eds.), *Organizational Constraints on the Dynamics of Evolution.* Manchester University Press, New York, pp. 279–290.

Szostak, J.W., Orr-Weaver, T.L., Rothstein, R.J., and Stahl, F.W. (1983). The double-strand-break repair model for recombination. *Cell* 33, 25–35.

Templeton, A. (1982). Adaptation and the integration of evolutionary forces. *In:* Milkman, R. (ed.), *Perspectives on Evolution.* Sinauer, Sunderland, MA, pp. 15–31.

Trivers, R.L. (1971). The evolution of reciprocal altruism. *Quart. Rev. Biol.* 46, 35–57.

Trivers, R.L. (1972). Parental investment and sexual selection. *In:* Campbell, B. (ed), *Sexual Selection and the Descent of Man.* Heinemann, London, pp. 136–179.

Uyenoyama, M.K. and Feldman, M.W. (1984). Theories of kin and group selection: a population genetics perspective. *Theor. Popul. Biol.* 38, 87–102.

Uyenoyama, M.K., and Waller, D.M. (1991a). Coevolution of self-fertilization and inbreeding depression. II. Symmetric overdominance in viability. *Theor. Popul. Biol.* 40, 47–77.

Uyenoyama, M.K., and Waller, D.M. (1991b). Coevolution of self-fertilization and inbreeding depression. III. Homozygous lethal mutations at multiple loci. *Theor. Popul. Biol.* 40, 173–210.

Uyenoyama, M.K. and Waller, D.M. (1991c). Coevolution of self-fertilization and inbreeding depression. I. Genetic modification in response to mutation-selection balance at one and two loci. *Theor. Popul. Biol.* 40, 14–46.

Valero, M., Richerd S., Perrot, V., and Destombe, C. (1993). Evolution of alternation of haploid and diploid phases in life cycles. *Trends Ecol. Evol.* 7, 25–29.

Van Roode, J.H.G., and Orgel, L.E. (1980). Template-directed synthesis of oligoguanylates in the presence of metal ions. *J. Mol. Biol.* 144, 579–585.

Van Voorhies, W.A. (1992). Production of sperm reduces nematode lifespan. *Nature* 360, 456–458.

Vrijenhoek, R.C. (1979). Factors affecting clonal diversity and coexistence. *Am. Zool.* 19, 787–797.

Wade, M.J. (1978). A critical review of models of group selection. *Quart. Rev. Biol.* 53, 101–114.

Walker, I. (1978). The evolution of sexual reproduction as a repair mechanism. Part I. A model for self-repair and its biological implications. *Acta Biotheoretica* 27, 133–158.

Weismann, A. (1889). *Essays upon Heredity and Kindred Biological Problems.* Clarendon, Oxford.

Whitehouse, H.L.K. (1982). *Genetic Recombination.* Wiley, New York.

Whittinghill, M. and Lewis, B.M. (1961). Clustered crossovers from male Drosophila raised on formaldehyde media. *Genetics* 46, 459–462.

Williams, G.C. (1957). Pleiotropy, natural selection and the evolution of senescence. *Evolution* 11, 398–411.

Williams, G.C. (1975). *Sex and Evolution.* Princeton University Press, Princeton.

Williams, G.C. and Mitton, J.B. (1973). Why produce sexually? *J. Theor. Biol.* 39, 545–554.

Wilson, D.S., (1975). A theory of group selection. *Proc. Natl. Acad. Sci. USA* 72, 143–146.

Wilson, D.S. (1980). *The Natural Selection of Populations and Communities* Benjamin/Cummings, Menlo Park, CA.

Woese, C.R. (1967). *The Genetic Code: The Molecular Basis for Genetic Expression.* Harper & Row, New York.

Woese, C.R. (1980). An alternative to the Oparin view of the primeval sequence. *In:* Halvorson, H.O. and Van Holde, K.E. (eds.), *The Origins of Life and Evolution.* Liss, New York, pp. 77–86.

Wojciechowski, M.F., Hoelzer, M.A., and Michod, R.E. (1989). DNA repair and the evolution of transformation in the bacterium *Bacillus subtilis.* II: Role of inducible repair. *Genetics* 121, 411–422.

Wojciechowski, M.F. (1992). DNA transformation, mechanism. *In:* Lederberg, J. (ed.), *The Encyclopedia of Microbiology.* Academic Press, San Diego.

Wurgler, F.E. (1991). Effects of chemical and physical agents on recombination events in cells of the germ line of male and female *Drosophila melanogaster. Mutat. Res.* 250, 275–290.

Yasbin, R.E., Wilson, G.A., and Young, F.E. (1975). Transformation and transfection in lysogenic strains of *Bacillus subtilis:* evidence for selective induction of prophage in competent cells. *J. Bacteriol.* 121, 296–304.

Zahavi, A. (1975). Mate selection—a selection for a handicap. *J. Theor. Biol.* 53, 205–214.

Zahavi, A. (1977). The cost of honesty (further remarks on the handicap principle). *J. Theor. Biol.* 67, 603–605.

Zahavi, A. (1978). Decorative patterns and the evolution of art. *New Scientist* 19, 182–184.

Ziehe, M., and Roberds, J. (1989). Inbreeding depression due to overdominance in partially self-fertilizing plant populations. *Genetics* 121, 861–868.

Index*